UNLIMITED PLAYERS

UNLIMITED PLAYERS

The Intersections of Writing Center and Game Studies

EDITED BY
HOLLY RYAN AND STEPHANIE VIE

UTAH STATE UNIVERSITY PRESS
Logan

Published by Utah State University Press
An imprint of University Press of Colorado
245 Century Circle, Suite 202
Louisville, Colorado 80027

The University Press of Colorado is a proud member of the Association of University Presses.

The University Press of Colorado is a cooperative publishing enterprise supported, in part, by Adams State University, Colorado State University, Fort Lewis College, Metropolitan State University of Denver, University of Alaska Fairbanks, University of Colorado, University of Denver, University of Northern Colorado, University of Wyoming, Utah State University, and Western Colorado University.

∞ This paper meets the requirements of the ANSI/NISO Z39.48-1992 (Permanence of Paper).

ISBN: 978-1-64642-193-0 (paperback)
ISBN: 978-1-64642-194-7 (ebook)
https://doi.org/10.7330/9781646421947

Library of Congress Cataloging-in-Publication Data

Names: Ryan, Holly Lynn, editor. | Vie, Stephanie, editor.
Title: Unlimited players : the intersections of writing center and game studies / Holly Ryan and Stephanie Vie.
Description: Louisville : University Press of Colorado, [2021] | Includes bibliographical references and index.
Identifiers: LCCN 2021036990 (print) | LCCN 2021036991 (ebook) | ISBN 9781646421930 (paperback) | ISBN 9781646421947 (ebook)
Subjects: LCSH: Writing centers—Administration. | Games—Study and teaching (Higher) | Game theory—Study and teaching (Higher) | English language—Rhetoric—Study and teaching (Higher) | English language—Composition and exercises—Computer-assisted instruction. | Tutors and tutoring—Study and teaching (Higher)
Classification: LCC PE1404 .U55 2021 (print) | LCC PE1404 (ebook) | DDC 808/.0420711—dc23
LC record available at https://lccn.loc.gov/2021036990
LC ebook record available at https://lccn.loc.gov/2021036991

CONTENTS

ACKNOWLEDGMENTS

HOLLY

I couldn't have done this project without the support of so many. First, I need to thank Stephanie Vie (affectionately known as S. Ellen), my co-editor, who has been an invaluable mentor to me during this project. Her experience, kindness, unbelievable organizational skills, and overall amazingness have made this project possible. I heart you, S. Ellen. A huge shout out to the Mid-Atlantic Writing Centers Association and all of its members for engaging and supporting the 2017 conference. Without that conference, this book would never have existed. I especially want to thank Margaret Ervin and Kelsey Hixon-Bowles for creating the conference CFP with me; their ideas deepened the call and I am grateful to them. Thank you to Penn State Berks for awarding me a research development grant to work on this book. Faculty need time to write and edit, and the grant afforded me that time (especially during the pandemic). It almost goes without saying that I am beyond grateful to the contributors to this collection; without their hard work, we wouldn't have a collection. Thank you to my writing center family at Penn State Berks who have listened to me talk about this collection for the past two years and who were as excited as I was when the collection was accepted for publication. Finally, thank you to my family, especially my husband, Patrick, who always encourages my work and believes in me. I couldn't ask for more.

STEPHANIE

Of course, I have to acknowledge my awesome co-editor, Holly "H. Lynn" Ryan. Working with her on this project has been an excellent experience, and I thank her for the many, many Skype and Zoom calls required to complete this edited collection during a pandemic. Thank you also to the University of Hawai'i at Mānoa's Outreach College, helmed by Dean William Chismar; I appreciate Outreach College's support of this project that allowed it to come to fruition. We thank all of the authors who contributed chapters to this collection and are excited to feature their work here in *Unlimited Players.* I hope their chapters and the games included will invigorate the work of writing and multiliteracy centers nationwide. I especially want to thank the collective knowledge and support of those in the Games, Culture, and Play track at the Southwest Popular/American Culture Association Conference, particularly Ken McAllister and Judd Ruggill, who gave us space to share this project at the annual SWPACA conference in 2018. Thanks for always being there to cheer on games-related scholarship and providing a supportive environment to play games and theorize them, too. And thank you to Rachael Levay and Darrin Pratt at University Press of Colorado / Utah State University Press, who believed in the value of this collection and helped shepherd us through the entire process (again, during a pandemic). Several cats were critical to the publication of this book: Mephistopheles Caligula Cthulu, Jalep(No!), and Ashley Katniss von Striperdeen. Every book should include the support of multiple cats. Last but not least, many thanks to Jeff Stockberger, who understands that sometimes I have to go to bed very late because academic writing and editing must happen. You're the best.

UNLIMITED PLAYERS

Introduction

WHY GAMES?

Toward a Theory of Gameful Writing Center Pedagogy

Stephanie Vie and Holly Ryan

This project began, as many do, with a favor. In this case, Holly asked Stephanie to be the keynote speaker at the Mid-Atlantic Writing Centers Association Conference she was hosting at Pennsylvania State University, Berks in Reading, Pennsylvania. The theme of the conference, A Day at the Carnival: Writing Centers as Sites of Play, begged for a keynote speaker who could theorize games and ignite a conversation about how writing centers could intersect with game and play scholarship. Stephanie's thoughtful keynote, "*Pokémon Go* Is R.A.D.: How Game Studies and Writing Center Research Can Learn from Each Other," asked questions about what writing center practitioners could learn from the study of augmented reality games and what connections there might be between writing center work and game studies research. Dynamic conversations emerged during this conference and from subsequent roundtable presentations by Holly and Stephanie at the International Writing Centers Association Conference in 2017 and the Southwest Popular/American Culture Association Conference in 2018. Through these conversations, we unsurprisingly learned that there were other scholars who inhabited the space between game studies and writing center studies but that very little scholarship worked to bring these two disciplines together. We want to fill that gap.

Therefore, in this collection, we work to bring together the fields of writing center studies and game studies. In doing so, we address several important research questions. First and perhaps foremost, we articulate the reasons why this overlap is productive. We start to answer the question: What does bringing together two seemingly disparate fields of study offer scholars who dwell in those overlaps? Second, we drill down to some of the specifics of this productive overlap; by doing so, we offer theoretically informed practices that writing center directors,

https://doi.org/10.7330/9781646421947.c000

consultants, staff, and others associated with writing centers can take away and apply in their own work. We work toward an answer to the question: What does a theory of writing center pedagogy look like when informed by game studies? We also hope to spur the thinking of game studies scholars and help build productive bridges between their work and that of writing centers. Writing studies as a field embraces game studies approaches, visible in the use of games in the classroom and game studies terminology and terministic screens in scholarship. We see many opportunities for those who work in writing centers to embrace game studies approaches, too. Thus, similarly and finally, we hope this collection will bring that enthusiasm for games and the study of games to writing centers, and vice versa: to help game studies scholars in writing studies and writing centers begin to question, What if?

What if we used games in the writing center and published about the impacts of doing so? What if we deepened already ongoing discussions about play and creativity in the writing center by incorporating language and theories from game studies? What if we made spaces for greater collaboration between game studies researchers and writing center practitioners? What if we used concepts and ideas from game studies to help writing centers grapple with the changing landscape of twenty-first-century writing and composing practices—practices that, as we explore momentarily, ask writing centers and their staff to engage with new media forms, digital compositions, multimodal writing, and the like? In short, what if games and gaming terminology became more familiar in the everyday practices and scholarship of writing centers? What would our work look like then? What would it mean to have an unlimited number of players coming together to collaborate on the best ways to educate tutors and work with writers? Instead of limiting the people in the game, what would happen if we open up writing center work to more diverse perspectives?

In this introduction, we first define our terms; specifically, we explain what we mean by games, play, multimodality, and new media. Then, we describe how attending to games in writing centers can offer new approaches to working with multimodality as well as new approaches to interdisciplinarity. We also explore how attending to games can productively deepen already existing conversations regarding the role of creativity, play, and engagement in writing center work. Following that discussion, we review the existing literature on the relationship between writing studies and game studies in an effort not only to model how these two fields inform one another but also to describe some of the theories and practices that unite writing and gaming. Finally, we describe the structure of this book and preview the arguments presented in each chapter.

DEFINING OUR TERMS

What Is a Game?

While game and composition scholars have been defining and operationalizing the value of play in their own intellectual spaces, writing center practitioners have a limited engagement with games. There are few references to play in writing center scholarship. Neal Lerner (2009) cites Helen Parkhurst's 1922 text as evidence of early laboratory writing practices. She writes, "The important thing is not to make young children study the thing they don't like, for the moment school is not as interesting as play it is an injury" (cited in Lerner 2009, 17). For Parkhurst and her contemporaries such as Thomas Nash (1984), who likens the writing center to a playground, writing centers allow for unstructured exploration of texts that mimic the world of playful imagination children inhabit when they are not in school. However, not until sixty-five years after Parkhurst, when Daniel Lochman (1986) wrote "Play and Game: Implications for the Writing Center," did writing center scholarship receive a sustained discussion of play, games, and their relationship to one another. For Lochman, play is liberating and unstructured, and, if accessed appropriately, it can "generate significant associations, imaginative insight, bold expression and valuable ideas" (14). Conversely, for Lochman, games discipline unbridled play and are defined by their "pursuit and acquisition of a goal" (14). He defines the relationship when he writes, "Together, play and game offer potential for the acquisition and communication of knowledge, since the undisciplined materials generated during play may be presented to an audience through the conventional, normative modes of expression appropriate to the game" (14).

We agree with Lochman that play is an unstructured, free association space for generative learning, but, as rhetoricians and writing scholars, it is challenging for us to imagine a context that is not constrained by some rules of engagement, those that are generated by a teacher, world builder, creator, or genre and audience expectations. For us, there are always rules to play by, even if those rules are defined by language constructs. Therefore, in this collection, we are interested in games and how gaming contexts intersect with writing, writing centers, and tutoring. Several of the authors in this collection break down the barriers between play/playfulness and games in writing centers in productive ways. An understanding of play is necessary to understand games, and vice versa.

Lochman begins to provide a workable definition of games (play that is defined by the pursuit of a goal), but, for our purposes, game scholars Katie Salen and Eric Zimmerman (2004) offer a richer way of defining games: "A game is a system in which players engage in an artificial conflict, defined by rules, that results in a quantifiable outcome" (68). Their definitions (see table 0.1) can be expanded upon and applied to writing center examples. Although there are many ways of applying Salen and Zimmerman's definition, our example attempts to orient readers toward thinking of writing center work as a game, much in the way Lochman does in his article. Table 0.1 presents Salen and Zimmerman's definitions and our examples.

While Salen and Zimmerman provide a broad definition of games, which serves as an effective base for our conversations, the attributes of games can be varied. For example, some might also say that games are "ethical technologies, capable of embodying values and projecting them into the user experience" (Sicart 2012, 101) since they are designed, developed, and created by and for humans, with all the values associated with and embedded into the game as a result of that design process. In the subsequent chapters of this collection, the authors often add on to this basic definition to enrich our understanding of games and play. For example, Elliott Freeman's chapter draws on Caillois's work to deepen our understanding of play, and Brenta Blevins and Lindsay A. Sabatino's chapter uses the concepts *ludus* and *paidia* to establish the framework for how play and games can inform an emergent theory of tutoring.

What Is Multimodality? What Is Multimedia? What Is New Media?

Three other key terms drive our collection: multimodal, multimedia, and new media. Others have done the historical work of tracing these terms (see Lutkewitte 2014; Palmeri 2012), which we do not intend to repeat here. Instead, we will provide working definitions of these terms from recent composition and writing center scholarship that can provide a grounding for our use of these terms. For us, multimodal is best defined by Pamela Takayoshi and Cynthia L. Selfe (2007) when they write that multimodal texts are "texts that exceed the alphabetic and may include still and moving images, animations, color, words, music and sound" (1). These texts are necessarily multimedia texts since they merge different kinds of media. Throughout this collection, when authors refer to multimedia, they are frequently referring to digital multimedia texts. Indeed, much of the game studies scholarship today focuses on digital games: video games, mobile and app-based games, and computer games.

Table 0.1. Mapping Salen and Zimmerman's definitions of game terminology onto writing center examples

Term	*Salen and Zimmerman (2004) Definition*	*Writing Center Example*
System	"A set of parts that interrelate to form a complex whole" (68).	The tutoring session is a system made up of a set of parts (e.g., the draft, the table, the computer, the assignment, the writing tools, the agenda setting, the tutoring strategies, the dialogue, the client report).
Player	"A game is something that one or more participants actively play. Players interact with the system of a game in order to experience the play of the game" (93).	At least one tutor and a writer make up the players.
Artificial	"Games maintain a boundary from so-called 'real life' in both time and space. Although games obviously occur within the real world, artificiality is one of their defining features" (93).	Since the conflict is defined by the participants (for example, writing the introduction to a paper during the time allotted), then it is artificial: both the time constraint and the goal are agreed upon by the people in the session.
Conflict	"All games embody a contest of powers. The contest can take many forms, from cooperation to competition, from solo conflict within a game system to multiplayer social conflict. Conflict is central to games" (93).	The tutor and the writer work cooperatively to overcome a problem during a specific session time frame (e.g., writing the introduction of a paper, which would have been a goal set out at the agenda-setting stage of the session).
Quantifiable Outcome	"Games have a quantifiable goal or outcome. At the conclusion of a game, a player has either won or lost or received some kind of numerical score. A quantifiable outcome is what usually distinguishes a game from less formal play activities" (93).	The reward can be extrinsic (e.g., accomplishing a goal established at the beginning of a session) or intrinsic (e.g., increased sense of self).

However, we note, too, that multimodality and multimedia texts may be non-digital in form (see Shipka 2011 for a discussion of multimodality beyond the digital). Several of the chapters in this collection explore analog games, such as fantasy RPGs (role-playing games) that can be played with dice, boards, or character sheets. Similarly, writing center tutors may use a range of physical tools to help a writer invent, perhaps even using mixed media such as paint, markers, cut-out shapes, and so on to work through the composing process; consultants and authors may frequently rely on computers to digitally compose word-processed documents, presentation slides, podcasts, and other forms. In this collection, we may use the terms *multimodal* and *multimedia* interchangeably, as multimodal compositions are inherently multimedia.

Finally, the term *new media* has been used in writing center scholarship and needs our attention. Two recent collections employ the term similarly. First, in 2010, David Michael Sheridan and James A. Inman co-edited *Multiliteracy Centers: Writing Center Work, New Media, and Multimodal Rhetoric.* In the introduction, Sheridan writes that new media takes various forms such as "web pages, digital video, and digital animations" (2). Sohui Lee and Russell G. Carpenter's (2014) definition includes Sheridan's examples (and also includes texts like PowerPoint presentations, electronic portfolios, and digital ethnographies), but they go much further in their definition in *The Routledge Reader on Writing Centers and New Media.* They define new media as "the cultural objects that . . . use digital technologies for distribution of information, communication, and data. [New media] encompasses the digital data and communication—from video to applications (apps) on cell phones . . . It means that consumers are also producers who can create, collaborate, and share content" (xvii). While we are slightly put off by the term *new media* (what is considered new, and at what point does something new become old, after all?), Lee and Carpenter's definition is effective for much of the work in this collection. Digital games are new media, and therefore online RPGs, virtual and augmented reality games, and even online tutoring would fall within the category of new media, whereas analog games such as in-person RPGs and tabletop games would not be a fit. Therefore, in our collection, the writers tend not to use new media as a term for their work, tending to use multimodal or multimedia instead, and again, several chapters explore analog but multimodal games such as RPGs (see LeCluyse, Shay and Shay, and Henthorn for examples in this collection). These terms—multimodal, multimedia, and new media—are slippery, and for our purposes in this collection, this is how we have chosen to approach them.

NEW APPROACHES TO LITERACY AND MULTIMODALITY IN THE WRITING CENTER

As contributors outline throughout the collection, one reason the overlap between game studies and writing center scholarship is productive for writing center studies is that it provides writing center practitioners with new or improved approaches to thinking about multimodality. For quite some time, writing centers have been concerned with the ways technology—computers specifically—has impacted tutoring practices on a range of compositions (Carino 2001; Grutsch McKinney 2010; Harris and Pemberton 1995; Hewett 2010; Lee and Carpenter 2014;

Pemberton 2003; Sabatino and Fallon 2019; Sheridan and Inman 2010; Trimbur 2010). Given writing center scholars' interest in multimodal composing and tutoring, engaging with games feels like the next logical step in the quest to understand these composing and tutoring practices. As noted earlier, games, both digital and analog, are multimodal texts; as such, they are yet another curve in a multimodal turn, a turn Jason Palmeri (2012) has illustrated is long-standing in writing studies.

From a 1995 special issue of *Computers and Composition* featuring articles on "writing centers online" (Kinkead and Hult 1995) to a 2016 special issue over two decades later on "pedagogies of multimodality and the future of multiliteracy centers" (Carpenter and Lee 2016), from edited collections on multiliteracy centers (Sheridan and Inman 2010) to collections on new media and writing centers (Lee and Carpenter 2014), writing center studies has long been curious about the impact of technological developments relevant to composing on writing center work. Michael A. Pemberton (2003) asserted that computers have been part of writing center work for the better part of forty years. They have specifically been used as tools with which to write, teach, or otherwise communicate; yet, as he noted, that relationship has been "only a cordial one, with occasional fluctuations ranging from wild enthusiasm to brooding antagonism" (11). He further cited Lerner (1998), Peter Carino (1998), Muriel Harris and Pemberton (1995), Nancy Maloney Grimm (1995), and others as early explorers of the possibilities and potential problems related to the incorporation of digital technologies in writing center work.

As digital and multimodal technologies began to impact the writing classroom and, by extension, writing center work, writing center scholars explored new media, multimodality, and multiliteracies with a heightened fervor. Jackie Grutsch McKinney (2009) not only outlined approaches to tutoring new media texts but also argued that writing center work needed to evolve to keep up with changing literacy practices. Andrea Lunsford and Lisa Ede (2011) similarly stated that "the growing importance of visual, oral, and performative rhetorics, not to mention of the digital revolution, has challenged us to extend our borders and expand our mission whenever possible . . . [This is] a key moment in writing center history, as writing becomes multimodal, multimedia, multilingual, and multivocal and as writing centers move to adapt to students' shifting communicative needs" (21). As these changing literacy practices began to exert influence on writing classrooms and writing centers alike, conversations deepened to incorporate, among other topics, the importance of considering disability, accessibility, and social justice in conjunction with multimodal texts (Hitt 2012; Naydan 2013) as well as the intersections of multimodality, multiliteracies, and identity politics (Ballingall 2013). Several of the authors in this collection grapple with identity studies, literacies, and multimodality through the lens of games, such as Elizabeth Caravella and Veronica Garrison-Joyner, who explore gameful design and its possibilities for more inclusive multiliteracy centers; Christopher LeCluyse, who examines the identity play that can occur in writing centers through the example of fantasy role-playing games; and Jessica Clements, who provides the example of a gaming ethnography as a means to encourage future writing center tutors to better empathize with tutees while also retaining an intersectional approach to identity.

Games studies has also wrestled with and attended to the challenges of changing literacy practices in writing. Scholars such as James Paul Gee (2008), John Alberti (2008), Jonathan Alexander (2009), Jennifer deWinter and Stephanie Vie (2008), and Gail E. Hawisher and Selfe (2007) each provide ways to understand the intersections of literacy and multimedia texts. Game studies language, terminology, terministic screens, and scholarship can be brought to bear on our work as writing center practitioners and scholars, and by doing so we may find promising avenues we can draw from as we work with the increasing presence of multimodal composing in our daily activities and our scholarship.

Similarly, games themselves are increasingly becoming the central object of focus in many classrooms worldwide, writing classrooms included. We describe later in this introduction how the increasing prevalence of digital games in everyday life has led to a concurrent increase in the use of games in writing classrooms. The scholars we cite have articulated how writing studies scholars have brought in games as pedagogical offerings, including to teach writing in many forms; as teachable moments regarding critical cultural concepts such as race, gender, sexuality, social status, disability, and so on; and as writing tools themselves, such as when faculty ask students to create their own games using technologies like Twine or Unity and others. As games become more common in everyday life, they have become more common in writing classrooms. And as a result, they are becoming more common in writing centers, too. Bringing a consideration of games into writing center studies is just one more way we attend to calls to adapt to students' changing needs as composers and writers.

We note here, too, that game studies scholarship is itself necessarily interdisciplinary; games and the study of games belong to no one field, and while game studies is now cemented as a field of study (with the attendant conferences, peer-reviewed academic journals, MA and PhD programs, and other markers of an established scholarly field), most who place themselves within game studies as a scholarly home come from a wide variety of academic backgrounds and rely on varied scholarly methods. Frans Mäyrä (2009) notes that while "game studies [has developed] a conceptual, theoretical, and methodological corpus of its own," the interdisciplinary nature of game studies provides "the potential of game studies as a radical, transformative form of scholarly practice" (313). Paul Martin (2018) further states that "scholars interested in understanding games benefit from knowing not only the achievements of their disciplinary colleagues, but also the work done in other areas of the campus, and even outside the university's walls" (introduction). In articulating the benefits of bringing game studies into different disciplines, Martin says, "A particular disciplinary perspective runs the danger of focusing on one layer or process to the neglect of others. Multiple perspectives can help" (n.p.). Thus, in the next section, we describe the potential power of bringing an interdisciplinary approach to writing center activity and scholarship through the application of game studies work.

NEW APPROACHES TO INTERDISCIPLINARITY IN THE WRITING CENTER

Writing center professionals often discuss interdisciplinarity in three ways. First, interdisciplinarity refers to educating tutors to work with writers from across the disciplines (Devet 2014), sometimes focusing on educating tutors in transferable strategies that can work equally well for tutoring science lab reports or art critiques or teaching them genre conventions for a range of papers they may encounter during a session. Other times the focus is on educating tutors specifically in the genre-specific writing they may encounter. Beyond pedagogy, writing center scholarship theorizes and describes the value of collaborating with faculty across campus by creating or strengthening ties with Writing across the Curriculum/Writing in the Disciplines (WAC/WID) programs (Arzt, Barnett, and Scoppetta 2009; Barnett and Blumner 1999; Harris 1992; Mullin 2011; Pemberton 1995). Finally, writing center practitioners as a whole are interdisciplinary, coming from a range of areas, often in the humanities or social sciences but not necessarily from English or rhetoric and composition. As well, writing center scholarship is also necessarily interdisciplinary, drawing from a range of theories and practices beyond the narrow scope of writing studies to better articulate and understand writing center praxis.

For writing center practitioners, especially students, an interdisciplinary approach to the everyday practices within the writing center can offer new terminology and new guidance for the work they do. Writing center work has long been welcoming to faculty of all ranks, both on and off the tenure track, and particularly welcoming to graduate and undergraduate students, given the prevalence of both in writing center tutoring and consultant positions. As with many overlaps between different fields, incorporating concepts, ideas, metaphors, theoretical lenses, and so on from another discipline offers value to scholars and practitioners in each area.

For example, undergraduate and graduate students who work as writing center consultants are frequently already familiar with video and computer games in a variety of forms but possibly lack the terminology and the scholarly apparatus necessary to theorize games in the writing center. Lee and Carpenter (2014) explained in *The Routledge Reader on Writing Centers and New Media* that their collection "acknowledges the many years of excellent writing center scholarship but also foregrounds the need for connecting our research with other fields that have explored how new media shapes communication" (xv). Similarly, we see our collection as continuing such an exploration, honing in on game studies—and thus games, which are multimodal texts—to investigate the ways games and play prompt us to re-envision writing center practices and conversations.

Explorations of multimodality in writing studies have necessarily drawn on interdisciplinary approaches; our collection's approach that brings together game studies and writing center studies is also necessarily interdisciplinary. Raymond C. Miller (1982) described interdisciplinary approaches as "all activities which juxtapose, apply, combine, synthesize, integrate or transcend parts of two or more disciplines" (6). Further, he articulated, each discipline shares its own worldview, the "underlying premises of thought" or the "conceptual construction which is used by a group to interpret reality" (5). Within writing and rhetoric, we might approach this concept of worldview through Kenneth Burke's language of terministic screens, "conceptual vocabularies used to name and interpret the world, which includes the material phenomena and forces studied by science as well as the products or insights of human relations and thought. Terministic screens consist of the words we use to represent reality, and as selections from among many conceptual vocabularies, they can lead to different conclusions as to what reality actually is" (cited in Blakesley 2017, 1745).

Burke's concept of terministic screens showed that "language—inherently metaphorical—constructs rather than reflects knowledge"

(Jay 1988, 355). Thus we see in this collection an opportunity to provide new terministic screens, new language, for scholars in writing centers and in games with which to conduct their work.

DEEPENING DISCUSSIONS OF CREATIVITY, PLAY, AND NOISE IN THE WRITING CENTER

Later in this introduction, we provide a literature review of scholarship in rhetoric and composition (sometimes also referred to as writing studies) that draws on game studies theories and terministic screens. Such scholarship illustrates that rhetoric and composition has had a growing interest in both theorizing and applying games in writing and in the classroom. Where we see the gap this collection fills, however, is in the lack of scholarship within writing center studies—itself an area of focus within rhetoric and composition more broadly—that attends to games. With a small number of exceptions, few scholars have taken up research (broadly understood) on games in the writing center.

This dearth seems odd to us: writing centers have long been spaces for playfulness, for play, and for games. Writing center scholars have embraced the role of creativity (Dvorak and Bruce 2008), play (Lochman 1986; Welch 1999), and activity (Boquet 2002) in the work of writing centers, but as we noted earlier, little work has discussed games in the writing center. While Lochman (1986) argued for the combination of rules and regulations alongside play for writing centers—noting, for instance, that "play with language must be restrained by rules and conventions if it is to communicate" (16)—his extended discussion of the value of play for writing center work and its explicit connection to games through the idea of the "game of academic writing" is one of the first lengthy conversations focusing on this topic.

Later, Kevin Dvorak and Shanti Bruce (2008) assembled a compendium filled with authors who traversed the many opportunities for play in the writing center: incorporating play and toys (Verbais 2008), using role playing and interactive performance (McGlaun 2008), including playfulness in tutor training (Zimmerman 2008), and others. The editors ascribe their purpose in assembling this collection as pushing back against the institutionalization of the field, noting that when a field becomes more established, it runs the risk of becoming "stale, institutional," and stagnant (xii). Dvorak and Bruce focus on collecting ways contributors incorporated creativity into their writing centers in an attempt to "prove . . . that writing centers can include creativity and serious play alongside serious work—or better still—can put play to work, seriously" (xiii).

But that word *serious* continued to undergird discussions of play and creativity in the writing center, with later scholars (much as Lochman, Dvorak, and Bruce had earlier) echoing the need for a balanced approach between play and seriousness in the writing center. For example, Dvorak and Jaimie Crawford (2017), in ending their chapter on cross-institutional collaborations and writing center pedagogy, specifically call out the role of play in their takeaways for those intending to enact writing studies pedagogies on their campuses. "Remember to play and to encourage play in the work environment," Dvorak and Crawford state, "but . . . understand the benefits and potential drawbacks of a playful environment" (128). Frequently, when play and games are mentioned, they are discussed within a framework that assumes their incorporation will bring along with it negative possibilities, thus illustrating one of the tensions that impacts both writing center and writing studies practitioners alike to this day. Even in a recent tutor's column in *Writing Lab Newsletter*, Amelia Hall (2016) wrestles with this tension, here in regard to the use of puns in the writing center: "Puns are typically thought to be antithetical to serious scholarly writing, but their potential usefulness, in combination with the evolving genres of academic discourse, brought me to this question: Is there room for a writer's words to be playful within a discipline, while still maintaining scholarly dignity?" (23). Hall's words showcase the fact that the binary of seriousness and playfulness undergirds writing center work even beyond the use of games or play. Perhaps Scott Miller (2008) describes this state of being best when he asserts it as living "in a contrary state, and we are conflicted. We want to have play, with all of its wild possibilities; but we're afraid of precisely those possibilities, and also afraid that the play will just make us even less relevant than we already are" (26).

That is, our conversations about play and playfulness and about games always acknowledge the sometimes unspoken (but sometimes articulated loudly) question, If we bring play/games into the writing center/writing classroom, will anyone take us seriously? And will students *really* learn? Writing centers, too, have grappled with this legitimacy as a field, and adding games to our already marginalized position might be questioned. Yet, as we outlined above, we see great value in encouraging more scholarship on games in writing center studies and in bringing games (not just play or playfulness but games specifically) into writing centers themselves. If we buy into this narrative of play being equated with frivolity and work (e.g., writing, writing tutoring) as being equated with seriousness and appropriateness, then what do we miss out on? Albert Rouzie (2000) argues that these "deeply entrenched divisions . . . ultimately impoverish our culture's approach to literacy" (628–29).

Too, game studies scholars have struggled with similar divisions between so-called serious games and games that are played simply for the purpose of leisure or entertainment; as several of the contributors to this collection explore, this division between games as serious objects and games as mere leisure—or the division between games with a "civilizing quality" (Caillois 1961, 27), what may be called *ludus*, and games with "diversion, turbulence, free improvisation, and carefree gaiety" (13), or *paidia*—has occupied the attention of many game scholars (see Freeman as well as Blevins and Sabatino in this collection for detailed explorations of *ludus* and *paidia*). We liken the back and forth of these divisions to the push and pull between "serious" writing center work and creative, playful, gameful writing center work. Rather than spend additional time occupying this binary, we propose to explode the division entirely by asking readers to accept that gaming can be, at different times, creative, playful, serious, educational, purposeful, frivolous, fun, and beneficial. Like Scott Miller (2008), we believe we can embody both ends of the binary pole at the same time (43). The simple act of incorporating games or play into a writing center or classroom space does not immediately mean that purposeful learning is occurring there; further, it does not mean that only levity and frivolity are now present.

Instead, we must consider what the goals are for incorporating games into an educational space such as a writing center and allow ourselves opportunities to embrace the full spectrum of possibility when games are included. Unlike other technologies, games have not yet been accepted to the point of becoming invisible technologies in educational contexts; as Selfe (1999) asserts, once technologies become invisible, we begin to assume their neutrality, allowing us to "focus on the theory and practice of language, the stuff [we believe is] of real intellectual and social concern" (413). In her foundational article, "Technology and Literacy: A Story about the Perils of Not Paying Attention" (1999), Selfe states that "in the case of computers—we have convinced ourselves that we and the students with whom we work are made of much finer stuff than the machine in our midst, and we are determined to maintain this state of affairs" (414). In the case of games, playfulness, and creativity, have many of us not similarly acquiesced that our work in writing centers and in education should be made of "much finer stuff," more serious stuff? Instead, we wish to resist that urge and instead bring the relationship between gaming and writing center work, following Elizabeth H. Boquet and Michele Eodice's (2008) call, "into the realm of conscious awareness and consideration" (4), the kind of "paying attention" that Selfe (1999) urges would put scholarship and research in the writing center as praxis (432). This collection is our attempt to do just that. Certainly, there is space for further attention to the role of play, playfulness, and creativity in the writing center, and we believe specifically that the role of games in particular is a rich yet undertheorized aspect of writing center work.

In the next section, we review literature from writing studies scholarship that focuses on games and their study. Our literature review outlines work that has already occurred in rhetoric and composition broadly to illustrate that writing studies has taken up and legitimized the study of games vis-à-vis the teaching and theorizing of writing. In doing so, it opens up further space for writing center work—as a subset of writing studies broadly—to also attend to games. The scholarship in the next section therefore points to a space where writing center studies scholars and practitioners might take up ongoing conversations and questions and add to this literature.

LITERATURE REVIEW OF GAMES AND WRITING SCHOLARSHIP

Unlike writing center scholarship, rhetoric and composition (or writing studies) scholarship broadly has paid far greater attention to the study of games. Games in rhetoric and composition are both objects of theory and objects of practice. That is, rhetoric and composition scholars who study games have frequently approached the subject with an eye toward games' pedagogical use. This is unsurprising given the field's focus on the daily work of writing instruction, and much of this scholarship has investigated the potential pedagogical power of games: for example, addressing topics such as how to teach technical communication genres like walkthroughs or encouraging students to apply themselves throughout a course by gamifying the writing classroom (Finseth 2015; Grouling et al. 2014; Roach 2015). A number of authors describe the use of existing games (e.g., *World of Warcraft*) to help students better understand writing and rhetorical principles (Colby and Colby 2008), while others talk readers through the process of developing their own games in-house for classroom use (Balzotti et al. 2017; Sheridan and Hart-Davidson 2008). Others still conduct original research on the impact of games in the writing classroom by bringing into their pedagogy projects such as gaming literacy narratives (Arduini 2018).

However, such scholarship does not remain solely at the level of addressing pedagogy. Indeed, many rhetoric and composition scholars who study games have moved beyond the classroom to instead consider the impact of games on the field itself and our approaches to critical constructs such as literacy, writing, and composition—all central to the identity of the field (or, to return to our earlier discussion of interdisciplinarity and terministic screens, all concepts that shape our worldviews as those who study writing). More broadly, some scholars in rhetoric and composition turn to the field itself when considering how writing studies and game studies might overlap. Rather than focus on individual games or gamified classrooms, or study one's own classroom using games, these scholars theorize the larger impact of games on the field of rhetoric and composition. While interest in new media and digital forms of multimodal composing had been of significant interest to the field prior to the entrance of game studies on the scene, it was not until the early 2000s that rhetoric and composition scholars began to significantly examine games and game studies. Some of this early work (circa 2000–2009) addresses the intersections of games and literacies, perhaps most famously in Gee's (2003) book *What Video Games Have to Teach Us about Learning and Literacy*; while Gee, a linguistics scholar, addresses the K–12 teaching environment rather than the post-secondary classroom in his book, it has been widely cited in rhetoric and composition because it touches on topics of interest such as transfer of knowledge and learning as a social activity.

Closer to home, rhetoric and composition scholars who draw on Gee's work during this early stage of scholarship on games in our field frequently consider games' impact on literacies (Alberti 2008; Alexander 2009; deWinter and Vie 2008; Hawisher and Selfe 2007). Others often use principles drawn from Gee's book as support for their own analyses within writing studies. See, for example, Zach Waggoner's (2009) *My Avatar, My Self: Identity in Video Role-Playing Games*, which uses Gee's principles of real-world identity, virtual-world identity, and projective identity as starting points for Waggoner's own "terminological continuua" (i.e., "virtual/non-virtual" and "verisimulacratude/verisimilitude") that could be used by rhetoric scholars to better study the rhetoric of video games and identity construction (1).

Much of the work around this time period appears in special issues of journals (Colby and Colby 2008; Johnson and Lacasa 2008) or in edited collections (Hawisher and Selfe 2007) and monographs (McAllister 2004; Waggoner 2009). These publications laid early foundational ground for the next wave of scholarship to come in 2010 and beyond. For instance, Ken McAllister's 2004 monograph *Game Work* offers a substantive theoretical framework drawing on rhetorical theory for those in rhetoric and composition to use when studying games. His "grammar of gamework" (44) gives scholars a five-part framework to study "how meaning may be made and managed specifically by those who design, market, and play computer games" and a means to "talk about the processes and techniques involved in this meaning-making process" (43).

Later, 2008 seems to have been a turning point in the field, a time when scholars began to take particular notice of games and investigate their possibilities in rhetoric and composition. While some earlier work on games and rhetoric and composition exists (see, for example, Derrick 1986), such work is sparse and far between. Instead, the scholarship did not truly pick up speed until the early 2000s. Ryan Moeller and Kim White (2008) describe that "the computers and composition community [is] rapidly accepting the idea that using computer games in the classroom can be a very effective way to teach writing. For example, the 2008 program for the Conference on College Composition and Communication included at least 23 presentations that either featured *games* as their primary topic or recognized that game theory has other applications in the composition classroom" ("Abstract").

At the same time, the conference game *C's the Day*—now familiar to many attendees of the Conference on College Composition and Communication (CCCC)—was emerging, first as a pre-conference workshop in 2009 that developed a preliminary game called *Confarganon* and later as the early iteration of what we now know as *C's the Day*, first seen at the 2011 CCCC (deWinter and Vie 2015, writing about "History of C's the Day"). These moments, which included a burgeoning number of publications on games in rhetoric and composition; conference presentations, special interest groups, and workshops related to games offered at major conferences in the field such as CCCC and Computers and Writing; and the development of *C's the Day* as an official conference game sponsored by CCCC and the National Council of Teachers of English, all seemed to announce that the study of games in the writing classroom and the field itself was surging. And such a surge seemed also to dovetail with increased calls, such as Kathleen Blake Yancey's (2004), in the field for greater attention to new forms of composing that would move the field beyond "print only" and "words on paper" (298).

Scholarship on games in rhetoric and composition continues to emerge, deepening the ongoing conversations established between 2000 and 2009. At the time we are writing this introduction, game-focused work in rhetoric and composition has expanded substantially to include studies of new games and transmedia storytelling, wearable technologies (Euteneuer 2018), and augmented reality games. Additional edited collections (Colby, Johnson, and Colby 2013; Eyman and Davis 2016) and monographs collect further research that addresses gaming in rhetoric and composition. In these varied topics and approaches, we see both an expansion and a breadth of the media to which the field attends as well as an elaboration and maturation of ongoing conversations around literacy, technology, and the role of writing and composing.

Despite pointed attention to game studies within rhetoric and composition as outlined above, tensions remain today between acceptance of this part of our field's work and rejection by those who embrace a more traditional, gatekeeping view of literacy and writing studies. The latter see little space for gaming (or indeed for many other digital and multimodal forms of composition) in writing studies and pedagogy. Rebekah Shultz Colby's 2017 study of the field and its incorporation of games clearly addresses this issue. Her study examines, through a survey and follow-up interviews, the frequency with which writing faculty used games in their classrooms and in what ways. While she finds that few writing instructors responded that they used games in their teaching, she also explicates multiple reasons for this. Pointing to the tension between gatekeeping approaches to writing courses and faculty who incorporate digital and multimodal approaches to writing, Colby (2017) asserts that "the fact that writing teachers used video games the least in their assignments underscores curricular tensions within rhetoric and composition about what to value and privilege in writing instruction and how much multimodal composing should be taught compared to traditional academic written genres" (57). In this collection, we draw a similar parallel to understand the dearth of game-focused scholarship in the writing center, noting that writing centers, too, have historically faced stigmatization regarding their status in the university; by fighting against such stigmatization, some writing center staff, faculty, and tutors may have deliberately moved away from games—given their marked nature as stereotypical objects of low culture.

Within rhetoric and composition studies, some of this scholarship on games—as we have attempted to encourage throughout this volume as well—has drawn theories and terminology from game studies into writing studies broadly. See, for example, Richard Colby and Rebekah Shultz Colby's (2008) article describing the writing classroom as a space that is like Johan Huizinga's (1955) "magic circle," or as Colby and Colby put it, "a space bounded by terms and class periods and defined by its own set of classroom rules and learning objectives" (303). Joshua

Daniel-Wariya (2016), too, draws on the terminology of the magic circle to offer a rhetorical theory of play for writing studies, noting that "theorizing play's rhetorical potential is critical to the traditional and ongoing goals of computers and writing teaching and research in general" (45). He argues that instructors should deliberately seek out composing mediums that lend themselves well to playful interactions as well as reflect on how play might shape particular materials and then incorporate those materials into their classrooms.

Finally, still others move past a pedagogical focus to simply analyze and address the impact of games in society today, taking a broader cultural studies approach to understand, for example, the role of knowledge acquisition in playing massively multiplayer online games like *World of Warcraft* (Alexander 2017). Samantha Blackmon and Daniel J. Terrell (2007) examine race as a factor in understanding representation in video games; through their analysis of *Grand Theft Auto: Vice City*, the authors explore racial diversity in a popular video game franchise and its implications for establishing or reinforcing stereotypes of behavior attributed to particular races. And Lee Sherlock (2013) explores queer sexuality in role-playing games such as *World of Warcraft* (*WoW*), using the central example of "a transgender player [who] was forcefully confronted by the hegemonic values surrounding gender and sexuality held by much of the *WoW* playerbase" (162) to point to the need for further interrogation of identity, sexuality, and online game play. These examples showcase the fact that writing studies scholars have approached games beyond classroom use and instead have focused on cultural and psychological studies.

One reason why we aim to bring game studies (through the lens of rhetoric and composition) and writing center studies together in this volume is to respond to calls such as that from Lee and Carpenter (2014), who ask writing center scholars to connect "our research with other fields that have explored how new media shapes communication" (xiv). Indeed, rhetoric and composition continually examines how new technologies impact writing and the teaching of writing, a clear indicator that the field is waking up to the fact that writing (and composing) today occurs frequently in digital contexts and with the aid of digital technologies. The burgeoning interest in rhetoric and composition in a narrower discussion of a particular set of digital contexts and technologies—that is, video games—is yet further proof that writing scholars are attending (as they should) to digital technologies with which their students are familiar. In this volume, we ask the question of writing center scholars: Why not games? We hope this volume will collect important voices from writing center studies who are interested in talking about games, play, and writing centers.

IN THIS BOOK

This collection is divided into three sections: Key Concepts, Terms, and Connections; Application of Games to the Writing Center; and Staff and Writing Center Education Games. The first two sections offer rich, theoretically informed substantial chapters that approach writing center work through the lens of games and play, and their contents cover a range of topics discussed in writing center scholarship: considerations of identity, empathy, and power; productive language play during tutoring sessions; writing center heuristics; and others. The final section includes games directors and tutors can play in the writing center. These games could be used for staff development but could also be played with writers to help them develop their skills and practices.

Part 1: Key Concepts, Terms, and Connections begins with Elliott Freeman's discussion of the writing center as a place for play. Drawing on concepts central to game studies work—*paidia* and *ludus*—Freeman skillfully applies these terms through Roger Caillois's (1961) four-part framework of play to expound on the powerful role play can offer to writing centers. Building on Beth Boquet's *Noise from the Writing Center* (2002), Freeman argues that "play is not just a way of creating noise or a by-product of it, but instead, play represents a powerful tool for channeling and modulating noise. If noise empowers and enlivens our work, then play provides us with a powerful mechanism for mindfully and knowingly directing that energy." In chapter 2, Neil Baird and Christopher L. Morrow tell the story of Libbie, a writing center tutor who overcame challenges in her consultant role through the careful application of game studies–based heuristics. Here, heuristics, a tool familiar to writing center practitioners, are updated through game studies concepts, and the authors believe the "dynamic nature of game studies heuristics—specifically the heuristic circle along with positional and directional heuristics—offers a new way of conceiving heuristics within the context of writing centers."

In chapter 3, Jason Custer offers a historical overview of process, another concept familiar to writing studies, to demonstrate that process in composition studies influenced procedurality in game studies; as he states, "seeing process across these fields presents an exigence for writing center practitioners and pedagogy to consider how focusing on concepts such as play and process may help students become better writers." The final chapter in this section, from Elizabeth Caravella and Veronica Garrison-Joyner, explores the work multiliteracy centers can do in "interrogating and dramatically restructuring the parameters of typical or conventional forms of multicultural discourse in writing center practice"; through making connections between policy and power explicit, tutors and students may be better empowered to experiment and play productively with language and mode. Gaming concepts and language such as procedural rhetoric, possibility spaces, and gameful design are used to sustain more inclusive and supportive writing center and multiliteracy center practices.

Part 2: Applications of Games to the Writing Center offers a set of six chapters that explore writing center practices and games, and many provide detailed takeaways for readers that we hope can infuse their own writing centers and practices. The first chapter in this section, from Brenta Blevins and Lindsay A. Sabatino, turns to augmented reality games—perhaps most familiar in the example of *Pokémon Go*—to provide an emergent theory of tutoring that better allows consultants to respond to multimodal writing in their centers; they close by stating that "just as players level up in highly re-playable emergent games with no clear termination, a tutoring theory of emergence recognizes that a written product isn't the ultimate objective, but instead, ongoing encounters with writing will yield new challenges, new opportunities for exploration, creativity, and discovery in new media." Next, in chapter 6 of the collection, Christopher LeCluyse draws together threshold concepts of writing and fantasy role-playing games to discuss how these two elements together can better equip writing consultants to serve students in sessions, "making the moves of both systems explicit . . . and creat[ing] a playful space to explore alternative subjectivities." Kevin J. Rutherford and Elizabeth Saur explore how the game studies concept of magic circles can emphasize inclusivity and equity in the writing center. The authors reflect on their experiences with opening a new university writing center and tie those experiences to the concept of the magic circle in an effort to illustrate their efforts to create a "more equitable and just game."

In the second half of part 2, Thomas "Buddy" Shay and Heather Shay also turn to role-playing games through the lens of dramaturgy, here applying the concept to writing center tutors and the three layers of identity they take on while working. Drawing on Dennis Waskul and Matt Lust's (2004) examination of the layers of identity among participants in tabletop role-playing games, Shay and Shay explain how tutor training can be modified to give consultants heuristics that equip them to navigate their multifaceted identities. The final two chapters in this section are explicitly pedagogical, offering readers assignments and activities that can be used in their writing centers. Chapter 9, from Jessica Clements, showcases a gaming ethnography assignment for tutor training that aims for greater intersectionality. And the final chapter, from Jamie Henthorn, reports on a pilot study that asked students in a tutor education course to participate in semester-long quests; by "turning her class into an RPG," Henthorn attends to how playfulness in consultant development offers greater ownership over their own professionalization.

Part 3: Staff and Writing Center Education Games offers ten practical games and playful activities that tutors, writing center professionals, and writers can play in the writing center and during staff development. The four staff education games are based on commercial games that may be familiar to our readers. For example, Nathalie Singh-Corcoran and Holly Ryan's game *Writing Center Snakes and Ladders*, which helps tutors discuss difficult tutoring scenarios, is based on the popular children's game *Chutes and Ladders.* Stacey Hoffer's *Active Listening Uno* and *Heads Up! Asking Questions and Building Vocabularies* are both based on card games. These two games provide tutors with a fun and interactive way to learn terms and concepts related to tutoring. Finally, Rachael Zeleny uses a familiar television show (*Shark Tank*) to frame her game "*And Now Presenting*: Marketing Writing Center Identities," in which writing center tutors create a marketing pitch for their writing center.

Other games in this section will be new to readers, such as the puzzle-based game by Christina Mastroeni, Malcolm Evans, and Richonda Fegins. In this game, tutors build group cohesion while also familiarizing themselves with tutoring resources. Two new games in the role-playing tradition are Alyssa Noch's *Level Up* and Mitchell Mulroy's "Writing and Role Playing," both of which encourage tutors to play with writer's identities, shifting the language of progress in a tutoring session to something more in line with role-playing discourse.

The final three games in the collection are intended to encourage discussion and open up conversations about issues related to tutoring. Elysse T. Meredith and Miriam E. Laufer describe a free-form non-synchronous play activity in which tutors write comments on a writing wall. Katie Levin's game "One Word Proverbs" similarly does not have a win-state but is a playful activity designed to encourage tutors to actively listen, collaborate, and find ways to write collaboratively. Brennan Thomas, Molly Fischer, and Jodi Kutzner take a play approach to citation styles with their game *Source Style Scramble.*

Each of the chapters in this collection answers the questions we posed at the beginning of our introduction: What would happen to writing center theory and practice if we brought together writing center scholarship and games scholarship? What would it mean to not limit our scholarship to familiar educational theories/theorists but to bring in other voices from scholars who typically are not part of writing center conversations? We would argue that opening the writing center scholarly play space to an unlimited number of players can bring about exciting and new ways of addressing the important practices, beliefs, and values of writing center pedagogy.

REFERENCES

Alberti, John. 2008. "The Game of Reading and Writing: How Video Games Reframe Our Understanding of Literacy." *Computers and Composition* 25, no. 3: 258–69.

Alexander, Jonathan. 2009. "Gaming, Student Literacies, and the Composition Classroom: Some Possibilities for Transformation." *College Composition and Communication* 61, no. 1: 35–63.

Alexander, Phill. 2017. "KNOWing How to Play: Gamer Knowledges and Knowledge Acquisition." *Computers and Composition* 44: 1–12.

Arduini, Tina. 2018. "Cyborg Gamers: Exploring the Effects of Digital Gaming on Multimodal Composition." *Computers and Composition* 48: 89–102.

Arzt, Judy, Kristine E. Barnett, and Jessyka Scoppetta. 2009. "Online Tutoring: A Symbiotic Relationship with Writing across the Curriculum Initiatives." *Across the Disciplines: A Journal of Language, Learning, and Academic Writing* 6. https://wac.colostate.edu/docs/atd/technologies/arztetal.pdf.

Ballingall, Timothy. 2013. "A Hybrid Discussion of Multiliteracy and Identity Politics." *Praxis: A Writing Center Journal* 11, no. 1. http://www.praxisuwc.com/ballingall-111.

Balzotti, Jonathan, Derek Hansen, Daniel Ebeling, and Lauren Fine. 2017. "Microcore: A Playable Case Study for Improving Adolescents' Argumentative Writing in a Workplace Context." *Proceedings of the 50th Hawaii International Conference on System Sciences.* doi:10.24251/HICSS.2017.013.

Barnett, Robert W., and Jacob S. Blumner. 1999. *Writing Centers and Writing across the Curriculum Programs: Building Interdisciplinary Partnerships.* Westport, CT: Greenwood Publishing Group.

Blackmon, Samantha, and Daniel J. Terrell. 2007. "Racing toward Representation: An Understanding of Racial Representation in Video Games." In *Gaming Lives in the Twenty-First Century,* edited by Cynthia L. Selfe, Gail E. Hawisher, and Derek Van Ittersum, 203–15. New York: Palgrave Macmillan.

Blakesley, David. 2017. "Terministic Screens." In *The SAGE Encyclopedia of Communication Research Methods,* edited by Mike Allen, 1745–48. Thousand Oaks, CA: Sage.

Boquet, Beth. 2002. *Noise from the Writing Center.* Logan: Utah State University Press.

Boquet, Elizabeth H., and Michele Eodice. 2008. "Creativity in the Writing Center: A Terrifying Conundrum." In *Creative Approaches to Writing Center Work,* edited by Kevin Dvorak and Shanti Bruce, 3–20. Cresskill, NJ: Hampton.

Caillois, Roger. 1961. *Man, Play, and Games.* New York: Free Press of Glencoe.

Carino, Peter. 1998. "Computers in the Writing Center: A Cautionary History." In *Wiring the Writing Center,* edited by Eric Hobson, 171–94. Logan: Utah State University Press.

Carino, Peter. 2001. "Writing Centers and Writing Programs: Local and Communal Politics." In *The Politics of Writing Centers*, edited by Jane Nelson and Kathy Evertz, 1–14. Portsmouth, NH: Boynton/Cook.

Carpenter, Russell, and Sohui Lee, eds. 2016. "Envisioning Future Pedagogies of Multiliteracy Centers." Special issue of *Computers and Composition* 41, no. 2. https://www.sciencedirect.com/journal/computers-and-composition/vol/41/suppl/C.

Colby, Rebekah Shultz. 2017. "Game-Based Pedagogy in the Writing Classroom." *Computers and Composition* 43: 55–72.

Colby, Richard, and Rebekah Shultz Colby, eds. 2008, September. "Reading Games." Special issue of Computers and Composition Online on writing studies and games. http://www.bgsu.edu/cconline/gaming_issue_2008/ed_welcome_gaming_2008.htm.

Colby, Richard, Matthew S.S. Johnson, and Rebekah Shultz Colby, eds. 2013. *Rhetoric/Composition/Play through Videogames: Reshaping Theory and Practice of Writing*. New York: Palgrave Macmillan.

Daniel-Wariya, Joshua. 2016. "A Language of Play: New Media's Possibility Spaces." *Computers and Composition* 40: 32–47.

Derrick, Thomas J. 1986. "Dosequis: An Interactive Game for Composition Students." *Computers and Composition* 3, no. 2: 40–53.

Devet, Bonnie. 2014. "Using Metagenre and Ecocomposition to Train Writing Center Tutors for Writing in the Disciplines." *Praxis: A Writing Center Journal* 11, no. 2: 1–7.

deWinter, Jennifer, and Stephanie Vie. 2008. "Press Enter to 'Say': Using *Second Life* to Teach Critical Media Literacy." *Computers and Composition* 25: 313–22.

deWinter, Jennifer, and Stephanie Vie. 2015. "Sparklegate: Gamification, Academic Gravitas, and the Infantilization of Play." *Kairos: A Journal of Rhetoric, Technology, and Pedagogy* 20, no. 1. http://kairos.technorhetoric.net/20.1/topoi/dewinter-vie/.

Dvorak, Kevin, and Shanti Bruce, eds. 2008. *Creative Approaches to Writing Center Work*. Cresskill, NJ: Hampton.

Dvorak, Kevin, and Jaimie Crawford. 2017. "Cross-Institutional Collaborations and Writing Studio Pedagogy." In *Writing Studio Pedagogy: Space, Place, and Rhetoric in Collaborative Environments*, edited by Matthew Kim and Rusty Carpenter, 111–29. Lanham, MD: Rowman and Littlefield.

Euteneuer, Jacob. 2018. "Conspicuous Computing: Gamified Bodies, Playful Composition, and the Monsters in Your Pocket." *Computers and Composition* 50: 53–65.

Eyman, Douglas, and Andrea D. Davis, eds. 2016. *Play/Write: Digital Rhetoric, Writing Games*. Anderson, SC: Parlor.

Finseth, Carly. 2015. "Theorycrafting the Classroom: Constructing the Introductory Technical Communication Course as a Game." *Journal of Technical Writing and Communication* 45, no. 3: 243–60. doi:10.1177/0047281615578846.

Gee, James Paul. 2003. *What Video Games Have to Teach Us about Learning and Literacy*. New York: Palgrave Macmillan.

Gee, James Paul. 2008. "Video Games and Embodiment." *Games and Culture* 3: 253–63.

Grimm, Nancy Maloney. 1995. "Computer Centers and Writing Centers: An Argument for Ballast." *Computers and Composition* 12: 323–29.

Grouling, Jennifer, Stephanie Hedge, Aly Schweigert, and Eva Grouling Snider. 2014. "Questing through Class: Gamification in the Professional Writing Classroom." In *Computer Games and Technical Communication: Critical Methods and Applications at the Intersection*, edited by Jennifer deWinter and Ryan Moeller, 265–82. London: Routledge.

Grutsch McKinney, Jackie. 2009. "New Media Matters: Tutoring in the Late Age of Print." *Writing Center Journal* 29: 28–51.

Grutsch McKinney, Jackie. 2010. "New Media (R)evolution: Multiple Models for Multiliteracies." In *Multiliteracy Centers: Writing Center Work, New Media, and Multimodal Rhetoric*, edited by David M. Sheridan and James A. Inman, 207–33. Cresskill, NJ: Hampton.

Hall, Amelia. 2016. "Playfulness in Discipline: How Punning Witticism Is Transforming Criticism in the Writing Center." *Writing Lab Newsletter* 40: 23–26.

Harris, Muriel. 1992. "The Writing Center and Tutoring in WAC Programs." In *Writing across the Curriculum: A Guide to Developing Programs*, edited by Susan McLeod and Margot Soven, 109–22. Newbury Park, CA: Sage.

Harris, Muriel, and Michael Pemberton. 1995. "Online Writing Labs (OWLs): A Taxonomy of Options and Issues." *Computers and Composition* 12: 145–59.

Hawisher, Gail E., and Cynthia Selfe, eds. 2007. *Gaming Lives in the Twenty-First Century: Literate Connections*. New York: Palgrave.

Hewett, Beth L. 2010. *The Online Writing Conference: A Guide for Teachers and Tutors*. Portsmouth, NH: Boynton/Cook.

Hitt, Allison. 2012. "Access for All: The Role of Dis/Ability in Multiliteracy Centers." *Praxis: A Writing Center Journal* 9, no. 2: 1–7.

Huizinga, John. 1955. *Homo Ludens: A Study of the Play-Element in Culture*. Boston: Beacon.

Jay, Paul. 1988. "Modernism, Postmodernism, and Critical Style: The Case of Burke and Derrida." *Genre* 21: 339–58.

Johnson, Matthew S.S., and Pilar Lacasa, eds. 2008. "Reading Games: Composition, Literacy, and Video Gaming." Special issue of *Computers and Composition* 25.

Kinkead, Joyce, and Christine A. Hult, eds. 1995. *Computers and Composition* 12, no. 2. Special issue on writing centers and digital technologies. https://www.sciencedirect.com/journal/computers-and-composition/vol/12/issue/2.

Lee, Sohui, and Russell G. Carpenter, eds. 2014. *The Routledge Reader on Writing Centers and New Media*. New York: Routledge/Taylor and Francis Group.

Lerner, Neal. 1998. "Drill Pads, Teaching Machines, and Programmed Texts: Origins of Instructional Technology in Writing Centers." In *Wiring the Writing Center*, edited by Eric Hobson, 119–36. Logan: Utah State University Press.

Lerner, Neal. 2009. *The Idea of a Writing Laboratory*. Carbondale: Southern Illinois University Press.

Lochman, Daniel. 1986. "Play and Game: Implications for the Writing Center." *Writing Center Journal* 7: 11–18.

Lunsford, Andrea, and Lisa Ede. 2011. "Reflections on Contemporary Currents in Writing Center Work." *Writing Center Journal* 31: 11–24.

Lutkewitte, Claire. 2014. *Multimodal Composition: A Critical Sourcebook*. Boston: Bedford/St. Martin's.

Martin, Paul. 2018. "The Intellectual Structure of Game Research." *Game Studies* 18. http://gamestudies.org/1801/articles/paul_martin.

Mäyrä, Frans. 2009. "Getting into the Game: Doing Multi-disciplinary Game Studies." In *The Video Game Theory Reader*, edited by Bernard Perron and Mark J.P. Wolf, 313–29. New York: Routledge.

McAllister, Ken. 2004. *Game Work: Language, Power, and Computer Game Culture*. Tuscaloosa: University of Alabama Press.

McGlaun, Sandee. 2008. "Putting the 'Play' Back into Role-Playing: Tutor Training through Interactive Performance." In *Creative Approaches to Writing Center Work*, edited by Kevin Dvorak and Shanti Bruce, 115–34. Cresskill, NJ: Hampton.

Miller, Raymond C. 1982. "Varieties of Interdisciplinary Approaches in the Social Sciences: A 1981 Overview." *Issues in Interdisciplinary Studies* 1: 1–37.

Miller, Scott. 2008. "And Then Everybody Jumped for Joy! (But Joy Didn't Like That, So She Left): Play in the Writing Center." In *Creative Approaches to Writing Center Work*, edited by Kevin Dvorak and Shanti Bruce, 21–48. Cresskill, NJ: Hampton.

Moeller, Ryan, and Kim White. 2008. "Enter the Game Factor: Putting Theory into Practice in the Design of Peer Factor." *Computers and Composition Online*. http://cconlinejournal.org/gaming_issue_2008/Moeller_White_Enter_the_game/index.html.

Mullin, Joan A. 2011. "Writing Centers and WAC." In *WAC for the New Millennium: Strategies for Continuing Writing-across-the-Curriculum Programs*, edited by Susan H. McLeod, Eric Miraglia, Margot Soven, and Christopher Thaiss, 179–99. Urbana, IL: National Council of Teachers of English.

Nash, Thomas. 1984. "Derrida's 'Play' and Prewriting for the Laboratory." In *Writing Centers: Theory and Administration*, edited by Gary A. Olson, 182–95. Urbana, IL: National Council of Teachers of English.

Naydan, Liliana. 2013. "Just Writing Center Work in the Digital Age: De Facto Multiliteracy Centers in Dialogue with Questions of Social Justice." *Praxis: A Writing Center Journal* 11: 1–7.

Palmeri, Jason. 2012. *Remixing Composition: A History of Multimodal Writing Pedagogy*. Carbondale: Southern Illinois University Press.

Pemberton, Michael A. 1995. "Rethinking the WAC/Writing Center Connection." *Writing Center Journal* 15: 116–33.

Pemberton, Michael A. 2003. "Planning for Hypertexts in the Writing Center . . . or Not." *Writing Center Journal* 24, no. 1: 9–24.

Roach, Danielle R. 2015. "Pedagogy at Play: Gamification and Gameful Design in the 21st-Century Writing Classroom." PhD dissertation, Old Dominion University, Roanoke, VA. https://digitalcommons.odu.edu/english_etds/7/.

Rouzie, Albert. 2000. "Beyond the Dialectic of Work and Play: A Serio-Ludic Rhetoric for Composition Studies." *Journal of Advanced Composition* 20: 627–58.

Sabatino, Lindsay A., and Brian Fallon, eds. 2019. *Multimodal Composing: Strategies for Twenty-First-Century Writing Consultations*. Logan: Utah State University Press.

Salen, Katie, and Eric Zimmerman. 2004. *Rules of Play: Game Design Fundamentals*. Cambridge, MA: MIT Press.

Selfe, Cynthia L. 1999. "Technology and Literacy: A Story about the Perils of Not Paying Attention." *College Composition and Communication* 50: 411–36.

Sheridan, David Michael, and William Hart-Davidson. 2008. "Just for Fun: Writing and Literacy Learning as Forms of Play." *Computers and Composition* 25: 323–40.

Sheridan, David Michael, and James A. Inman, eds. 2010. *Multiliteracy Centers: Writing Center Work, New Media, and Multimodal Rhetoric*. Cresskill, NJ: Hampton.

Sherlock, Lee. 2013. "What Happens in Goldshire Stays in Goldshire: Rhetorics of Queer Sexualities, Role-Playing, and Fandom in *World of Warcraft*." In *Rhetoric/Composition/Play through Video Games*, edited by Richard Colby, Matthew Johnson, and Rebekah Shultz Colby, 161–74. New York: Palgrave Macmillan.

Shipka, Jody. 2011. *Toward a Composition Made Whole*. Pittsburgh: University of Pittsburgh Press.

Sicart, Miguel. 2012. "Digital Games as Ethical Technologies." In *The Philosophy of Computer Games*, edited by John Richard Sageng, Hallvard J. Fossheim, and Tarjei Mandt Larsen, 101–24. New York: Springer.

Takayoshi, Pamela, and Cynthia L. Selfe. 2007. "Thinking about Multimodality." In *Multimodal Composition: Resources for Teachers*, edited by Cynthia L. Selfe, 1–12. Cresskill, NJ: Hampton.

Trimbur, John. 2010. "Multiliteracies, Social Futures, and Writing Centers." *Writing Center Journal* 30: 88–91.

Verbais, Chad. 2008. "Incorporating Play and Toys into the Writing Center." In *Creative Approaches to Writing Center Work*, edited by Kevin Dvorak and Shanti Bruce, 135–46. Cresskill, NJ: Hampton.

Waggoner, Zach. 2009. *My Avatar, My Self: Identity in Video Role-Playing Games*. Jefferson, NC: McFarland.

Waskul, Dennis, and Matt Lust. 2004. "Role-Playing and Playing Roles: The Person, Player, and Persona in Fantasy Role-Playing." *Symbolic Interaction* 27: 333–56.

Welch, Nancy. 1999. "Playing with Reality: Writing Centers after the Mirror Stage." *College Composition and Communication* 51, no. 1: 51–69.
Yancey, Kathleen Blake. 2004. "Made Not Only in Words: Composition in a New Key." *College Composition and Communication* 56: 297–328.
Zimmerelli, Lisa. 2008. "A Play about Play: Tutor Training for the Bored and Serious, in Three Acts." In *Creative Approaches to Writing Center Work*, edited by Kevin Dvorak and Shanti Bruce, 97–113. Cresskill, NJ: Hampton.

PART 1

Key Concepts, Terms, and Connections

1

PAIDIA-GOGY

Playing with Noise in the Writing Center

Elliott Freeman

In her book *Noise from the Writing Center*, Elizabeth Boquet (2002) makes a case for the role of noise in the writing center—a vitalizing, disruptive noise that embraces risk, improvisation, and tension. By championing noise in the writing center, Boquet (2002) advocates for a vision of writing center work that resists easy answers, scripted sessions, and practicality for its own sake. Indeed, she envisions this noise as productive chaos: "If the writing center is to function as an apparatus of educational transformation, that order must develop out of chaos, not through the elimination of it" (84). Throughout the book, she gestures toward play as a tool for creating and maintaining noise but never engages fully with the power that play and games have to enrich writing center pedagogy. She is not alone in using this metaphor of play; whether in passing or in detail, many writing center scholars have invoked play, and continue to do so, as a way to free the writers they work with from associating writing with drudgery.[1]

These calls to treat writing centers as places of play recognize that writing is already implicitly and inexorably linked with play. In the absence of play, creative writing—and here I mean writing that *creates* not just prose or poetry but any sort of new understanding, argument, or analysis—is either impossible or at least severely diminished. Play is a notoriously slippery term, and not enough work has been done to operationalize play as a critical lens for writing center praxis. What *is* play, and how can writing center practitioners understand the play they're asking writers and consultants to participate in?

To this end, this chapter will rely on sociologist Roger Caillois's framework for play, putting his concepts in conversation with writing center theory and practice. It will focus primarily on his concepts of *paidia* and *ludus*, which describe a tension between free-form and structured

https://doi.org/10.7330/9781646421947.c001

play that is based in large part on the presence (or absence) of rules.[2] To a lesser extent, this chapter will also define and apply Caillois's four types of play, primarily as additional lenses for understanding *paidia* and *ludus.* Taken together, these frameworks will be used to expand on Boquet's metaphor of noise, to demonstrate that play is not just a way of creating noise or a by-product of it but instead to argue that play represents a powerful tool for channeling and modulating noise. If noise empowers and enlivens writing center work, then play provides a powerful mechanism for mindfully and knowingly directing that energy.

METAPHORS OF PRACTICE

After a long day of helping students master the em dash—sovereign of all punctuation—I turn to play to help me decompress: video games, board games, card games, and especially tabletop role-playing games in the vein of industry behemoth *Dungeons & Dragons* (although my own tastes tend toward smaller, narrative-driven games, especially *Fate* and *Nobilis*). In these games, I often find myself in the role of the Game Master (GM). For the uninitiated, while other players usually embody a single character, the GM is tasked with managing the rest of the world, reacting to the players' choices and presenting them with interesting situations. In this role, GMs function as facilitators, weaving together disparate threads to help shepherd the group toward a satisfying experience. Doing so requires them to consider and reconcile many implicit and explicit factors, including player temperaments and interests, time constraints, group dynamics, formal rules, and more.[3]

As a hobby, it can feel suspiciously like my work in the writing center. Helping my players reconcile their spoken and unspoken expectations requires emotional and intellectual labor, as well as careful prioritization between higher- and lower-order concerns. At the highest level, GMs are generally expected to maintain consensus within the group on issues of genre, tropes, and conventions; at the lower order, they are typically charged with adjudicating game- and story-based minutiae, especially in parsing unclear or potentially problematic rules.

Tabletop role-playing games are also deeply rooted in language play. One common archetype among players is the "rules lawyer," who assiduously (or pedantically, perhaps even maliciously) parses the text of a rule, often to enforce a personal vision of truth or to tilt a situation in their favor. Some games and GMs encourage players to speak in character, improvising dialogue based on their chosen personae. In most games, groups work together to construct a narrative, describing their shared adventures and building consensus. The texts that define a game's rules often take joy in language play themselves, especially by introducing exotic vocabulary (*ichor* is a particular favorite, and the game *Exalted* was my first introduction to the word *autochthon*).

Tutoring and role playing: the two feel alike to me in a way that informs and empowers my pedagogy as a writing consultant, but I must admit that I am reflexively and intrinsically skeptical about what might be called *metaphors of practice*. Like other disciplines, writing center theory, criticism, and praxis are replete with these metaphors of practice, including the field's original (and ongoing) examination of what we call ourselves: writing centers. Writing labs. Writing studios. Andrea Lunsford (1991) offered up models for writing centers as garrets and storehouses; Elizabeth Busekrus (2017) examined the center-as-Burkean-parlor. We have been clinics (Carino 1992), bodegas (Wilson 2012), and hospitals (Feltenberger and Carr 2011). In creating and re-creating our professional and scholarly identities, the writing center community has always relied on scholarship that embeds us within rich metaphorical frameworks to communicate our philosophy and our praxis to our colleagues, consultants, writers, and communities. These metaphors of practice are essential and inevitable. Human cognition is rooted in metaphor, building both our vocabulary and our instincts based on similarities between unlike things (Lakoff and Johnson 1980). But on the micro-level, the moment-to-moment work of navigating a session with a writer, these metaphors can seem gauzy, abstract, and inaccessible.

Consultants often create their own metaphors of practice—sometimes successfully, sometimes not. One I see frequently from novice tutors is the exhortation for students to think of the comma as a breath, a pause. It's an analogy that many reach for because their understanding of the comma—what it is and how it works—is obscured by years of laboriously developed instinct. While potentially useful in the short term, the analogy is imperfect and prone to failure and confusion based on the writer's personal sense of rhythm and cadence. Sometimes, our metaphors are simply our own, born out of a constellation of unique experiences that align in ourselves; while they may be personally productive, sharing them can be as frustrating as explaining dream logic.

I feel the need to outline this ambivalence because I am working against it in this chapter; I see in Caillois's model of *paidia* and *ludus* a way of understanding and pacing a consulting session that seems well suited to guiding the decisions consultants face in the moment. Its value is not just in play being similar to the work we do but in the idea that moving between *paidia* and *ludus* allows a consultant to guide the writer by either introducing, enforcing, renegotiating, or removing certain rules and challenges. Navigating this framework can help give a session structure and energy by drawing on the engagement that springs forth so readily from play.

PAIDIA AND *LUDUS*

In *Les Jeux et les Hommes* (translated, albeit imperfectly, as *Man, Play, and Games*), Roger Caillois (1961) responded to earlier work by Johan Huizinga, refining and complicating Huizinga's definition of play. While Huizinga worked to sketch a definition of play and its role in the creation and maintenance of cultures, Caillois focuses instead on articulating a model for understanding the various kinds of play and their potential interactions. To this end, his model incorporates four types of play—competitive *agon*, chance-based *alea*, mind-altering *ilinx*, and the experience of *mimicry*. He further situates these four types within a framework of *paidia* and *ludus*, a systemic tension primarily concerned with the presence or absence of rules.

While some scholars translate *paidia* and *ludus* as playfulness and gamefulness, respectively (for example, Lucero et al. 2014), this can be problematic. Caillois's original title, *Les Jeux et les Hommes*, collapses play and games into a single word—*jeux*—which can be translated as either (or even as both). Throughout this chapter, I will primarily use *play* to encompass both aspects of the French *jeux* as playfulness and gamefulness. Where the distinction between the two is particularly important, I will refer specifically to *paidia* or *ludus*.

In his definition of *paidia*, Caillois (1961) describes it as "the spontaneous manifestation of the play instinct" (28), likening it to sudden, exuberant movement—skipping, dancing, jumping for joy, or shivering with anticipation. In this form, rules and structure are implicit and in a constant state of flux: when children play, whether with toys or one another, roles and rules are constantly renegotiated—what seemed like a game of cops and robbers can turn on a dime when one participant, allegedly shot by another, reveals that he had a forcefield all along.

Ludus, then, is the natural outcome of paidic play. Through negotiation, rules and tactics begin to crystallize: paper beats rock, knights move across the board in leaping L shapes, and you're not allowed to peek when you're it at the start of hide-and-go-seek. Even the most sensible rules are fundamentally arbitrary; faced with the infinite possible actions within an otherwise empty space, *ludus* "disciplines and enriches" (Caillois 1961, 29) the exuberance of *paidia* by imposing restrictions the player(s) must struggle against. As the possibility-space begins to narrow with the introduction of more and more restrictions, players are able to turn their energy and attention toward specific challenges and the skills required to overcome them. For Caillois, *ludus* is based on a desire for "gratuitous difficulty" (27), by which he means that ludic play requires obstacles to realize satisfaction. It is, in short, the pleasure of achievement.

While Caillois was primarily concerned with describing the way paidic play crystallizes into ludic play through a process of rule adoption and stabilization, play can move in the opposite direction as well. According to Graham H. Jensen (2013), Caillois "effectively precludes the possibility of movement between the two 'genres' [of *paidia* and *ludus*] and thus undermines the transformative and generative power of play, which is derived precisely from the point at which *paidia* and *ludus* necessarily intersect" (71). In essence, when ludic play is stripped of some of its rules, players are empowered to renegotiate the game and approach a familiar play-system from new, unusual, and potentially productive angles. *Ludus* enriches *paidia* through discipline, frameworks, and obstacles; in return, *paidia* may denude *ludus* of its difficulty and arbitrariness, leaving it fallow and wild for players to reclaim and reconceptualize.

PLAYFUL NOISE

Play brings *noise*. The word *play* itself may conjure memories of raucous children, shuffling cards, slot machine sirens, and the roar of a stadium. Noise celebrates. Noise interrupts. Noise is the sound of play piercing through banality, riotous and rapturous.

Boquet (2002) calls, forcefully and frequently, for noise in the writing center. She argues that noise is a powerful and multifaceted metaphor for the disruptive potential of writing center pedagogy. It represents deliberate and meaningful inefficiency (51), distortion and tension (68–69), risk (79), and the fulcrum between order and chaos (84). "Where is the pleasure?" she asks, "Where is the fun? Where is the place where the writer and respondent can enter into a groove for that session?" (71) She connects her metaphor of noise to music and then to play itself, likening play to improvisation (76). Her arguments weave together all of these metaphors into a cohesive and compelling call for writing centers to consider themselves "a liminal zone where chaos and order coexist" (84). Throughout *Noise from the Writing Center*, she advocates fiercely and adamantly for transformative pedagogy but leaves uncertain what that pedagogy might look like. This deliberate ambiguity is for practical as well as philosophical reasons: embracing chaos means resisting overly proscriptive solutions that might loop back on themselves, trapping consultants in a calcified pedagogy (71). The circumstances are too many and too unique: *this* campus, *this* writing center, *this* consultant, *this* writer, *this* session.

Play, however, seems like a natural and flexible response to the need for noise. While *paidia* may seem more immediately noisy than *ludus* (ask any elementary teacher who has ever overseen recess about how loud children at play can *really* be), but *ludus* can also be a source of both literal and metaphorical noise: the airhorns after a touchdown, the klaxon on a quiz show, or the *kiai* of a martial arts exposition. Inside the center, that noise may be less raucous, less obvious, but no less energizing.

When a student comes to our center for the first time, I often hear them rush to self-identify as bad writers, as *not*-writers. For many of them, writing is the worst kind of drudgework, and they come in expecting the session to be tiresome as well. These are the students who most need noise in their sessions, to shake away the plaque that's been built up on them, often from bad experiences in the past. So we play. We joke. I try my best to help them laugh at some point in the first five minutes, if only to set a tone that invites them into the experience of play—a technique that has grown increasingly important both to and for me, especially as I made the transition from peer tutor to professional. Being mindfully and deliberately playful is a way of recapturing that spirit of peer-to-peer consultation, a way of flattening the hierarchy that exists between me and the student writers I work with, even if only slightly.

This is *paidia* at its best: an invitation to the playground. The paidic instinct is to renegotiate, to build together, to trust our gut. In its simplest form, it means stepping away from and outside of the paper, sometimes even literally, to escape its confines. For many writers, trying to revise a text when they're looking at the paper itself is like trying to remodel a room when they're locked inside it—which is to say, claustrophobic and self-defeating. In response, *paidia* excuses certain rules and returns to creativity and improvisation. This can be seen in free writing and brainstorming but also in focusing on strength and flair in writing, drawing some swagger and fire out of the student by helping them give themselves permission to recognize how much they already know about their topic.

The ludic instinct, then, works to create (or make visible) certain challenges, obstacles, and rules for the writer to overcome. This can feel at odds with the nondirective pedagogy that centers much of writing center theory, but rules don't need be cages for writers—they can be jungle gyms instead, frameworks that give them unique opportunities for play. Writing center practitioners might (and I do) issue challenges. One common one I issue to students is to not simply cut down a paper but to choose a specific percentage (usually 10%). Like all ludic rules, the percentage is (mostly) arbitrary, but setting a specific goal transforms it into something the writer can achieve and, in doing so, turns it into something that can help them crystallize a new understanding of brevity. It transforms the work into a tower of precariously stacked Jenga blocks, creating a strange pleasure out of difficulty.

It is tempting to think of *paidia* and *ludus* as a simple binary or even a continuum, but paidic and ludic elements always coexist in experiences of play. Dan Dixon (2009) even argues that *paidia* and *ludus* are "two separate aesthetic qualities both present during the playing of games" (1). There is always some order to the chaos and always some chaos in the order. "Pure" *paidia*, free from any rules or expectations, is in fact an impossibility because players are always being influenced by "bodily impulse, well-developed skills, and environmental or technical limitations . . . that shape their interaction" (Henricks 2010, 179–80). *Paidia* and *ludus*, in many ways, represent and map to strategies and techniques that are already used in writing centers. Their value, then, isn't in their novelty so much as in their ability to help consultants organize these strategies and navigate between them. Regardless of whether *paidia* and *ludus* constitute a strict binary, they exert pressure on one another, and tutors can use this pressure by understanding when to help the student loosen or tighten their grip on the rules that bind them—from the assignment, from grammar, from style, from themselves.

FOUR TYPES OF PLAY

In addition to his model of *paidia* and *ludus*, Caillois also outlined four essential types of play experiences: *agon* (competition), *alea* (chance), *ilinx* (vertigo), and mimicry. While I am primarily concerned with *paidia* and *ludus*, a few interesting similarities can be drawn between writing center work and these four categories. Understanding them illuminates some ways in which play is already pervasive in the writing center.

Of the four, *agon* and *alea* are the most self-explanatory: if asked to name a game, most people would probably name an agonic game (chess, football, *Magic: The Gathering*) or an aleic one (poker, roulette, craps). Like *ludus*, the primary pleasure in *agon* is in overcoming challenges; while it refers to competition, that competition can be against other players or against the self. *Alea*, on the other hand, is a passive experience; it is the pleasure derived from suspense and uncertainty, whether the spin of a slot machine or the shuffling of cards. *Alea* is exclusive to ludic play because to create suspense, there must be preexisting rules for interpreting the results of rolled dice, dealt cards, or other devices. Without the strong framework of *ludus*, the player could not experience the thrill of knowing that a result—solid, incontrovertible, non-negotiable—was fast approaching.

Agon's role in the writing center is perhaps the easiest to see because students *do* compete—with other students, with hypotheticals, and most of all with themselves. In these situations, it can be particularly productive to turn to the metaphor of consultant-as-coach. In serving as coaches, the consultants can be conceptualized as standing "at the sidelines (not in the center of the playing field), offering encouragement and advice based on experience and training, while the player expends the needed effort to succeed" (Harris 1992, 110). Coaches, however, are not just sources of advice and encouragement but also curators of challenge—a good coach actively shepherds their players to greater levels of performance. In this way, coaching is both ludic and especially agonic, demonstrating an asymmetrical relationship in which the coach is a facilitator of challenge. In the writing center, embracing *agon*, like embracing *ludus*, may mean invoking the metaphor of the coach more deliberately, empowering consultants to help writers set their own meaningful standards.

Alea is less immediately relevant to work in writing centers but derives its power from a feeling both writers and tutors will be familiar with. For Caillois, games of chance are pleasurable because they have a moment between *doing* and *knowing*—when you roll the dice or pull the lever on a slot machine, there is a delay between your action and the outcome, and in that moment, you find yourself suspended in uncertainty. This feeling of suspension between action and outcome is familiar to any writer who has ever had to turn in a paper and wait for a grade.

Ilinx and mimicry are less self-apparent because they are more often seen as play rather than games. Just as *alea* is exclusive to ludic play, *ilinx*—the experience of vertigo, of altered consciousness—is exclusive to *paidia*. Caillois uses *ilinx* to describe experiences meant to excite and arouse, from spinning in place to riding a roller coaster. Experiences of *ilinx* might happen in the writing center, especially in times of heightened anxiety that are prone to make anyone feel a little dizzy, but these experiences are usually involuntary and unwanted, less a question of play than of panic.

Mimicry, however, is the most productive of all four types of play when applied to writing center work. We say that children play house, play doctor, play dead, play the fool—but children aren't alone in that kind of play. For novice consultants, writing center work can begin with mimicking the moves they see other tutors making (the all-too-common exhortation to "fake it 'til you make it");[4] at the same time, the writers who come to the writing center are often participating in mimicry of their own, trying to adapt a scholarly or artistic persona—often including the ever-elusive concept of "the academic voice."

Three or four times a year, I work with writers composing their first major disciplinary research papers who, in aiming for what they think academic voice should sound like, end up burying themselves under an avalanche of jargon. These sessions are a joy to me because these writers have committed themselves to doing what was asked of them: read closely, pay attention to how writing in this discipline sounds and feels. Learning how to adopt that disciplinary voice is difficult work that requires writers to absorb jargon, tone, tropes, and style.

Helping them dig their way out of the avalanche often means a little bit of role playing, with an emphasis on play. I bluster up a phrase beaten mercilessly with a thesaurus—my go-to being "never enumerate your unmatriculated poultry until they have fully gestated," a comically unwieldy translation of "don't count your chickens before they're hatched." I ask them to play along—to take something from their paper and turn it into awkward synonyms, and what follows is a chance to interrogate the mysterious line between writing that is complex and writing that is complicated, writing meant to illuminate and writing meant to either obscure or impress. But that first moment of play and irreverence is always the key, a way of giving a student permission to say that the emperor has no clothes.

To call it mimicry is not a value judgment. Although the word has connotations of rude or unsophisticated behavior, mimic play is simply about the experience of taking on another role, donning a mask or costume or even just a point of view. In many writing centers, consultants already ask writers to role play, even if only in small ways—to consider their work as if they were an audience member, or the faculty member who assigned it, or someone else entirely.

A GAME OF ESSAYS

If all writing is play, perhaps academic writing, in all its many forms and intents, may be better understood through the lens of play.[5] To do so may seem strange, particularly because when we think of games, we think first and foremost of fun. However, fun is an elusive and empty word; it can mean so many things that it means nothing at all. Is fun actually intrinsic to play and games? *Should* it be? If play isn't fun, is it still play? The answer for Miguel Sicart (2014) is that fun really doesn't matter: play "is pleasurable, but the pleasures it creates are not always submissive to enjoyment, happiness, or positive traits. Play can be pleasurable when it hurts, offends, challenges us and teases us, and even when we are not playing" (3). To bring out playfulness in writing center pedagogy, it is vital first to consider the ways play is manifest in the kind of writing students bring to the center.

For Caillois (1961), play has six essential characteristics, being an activity freely engaged in, separate from the ordinary world, uncertain in outcome, unproductive, governed by rules, and make-believe (9–10). Many of these apply clearly and neatly to academic writing in its many genres and forms: writing assignments are clearly governed by rules and often require some element of make-believe, forcing the writer to adapt a stance or position that may or may not fully align with their own. Certainly, writing is uncertain in outcome—first drafts doubly so.

It is harder to say, then, that academic writing is something in which students freely engage, but students (usually) come to the writing center of their own volition; even when they are required to do so, they are still free to decide whether to engage with writing center workers and to what extent. When I work with students who so readily declare themselves to be non-writers, it is clear that they see papers and projects as separate from the "real world" of their everyday lives; because I work primarily with healthcare students, paper assignments often feel more like boxes to be checked off instead of opportunities to explore what they know. Rather than fight this instinct, it may instead make it easier to draw them into play. Treating the paper as play can grant students license to see the paper as its own system, one that can be played with and manipulated in a way that is inherently engaging.

Is it possible, then, to *win*?[6] The question is troubling because for many of the writers we work with, winning might be synonymous with getting a good grade. Writing centers traditionally avoid engaging with the topic of grading for any number of important reasons[7]—not the least of which is reifying the center's place as "outside at least some of the exigencies of academic life" (Bergmann 2010, 160). Students do not, however, have the luxury of sidestepping the issue of grading because it is an indisputable influence on their decision to come to the writing center (or not). How, then, can we help them become productively playful?

Consider *performance*. When one can "win" at ludic play, it usually involves some assessment of how the player performed. Games (often) make these assessments easy: when you take your opponent's king or jump their last checker, you win. But not all ludic play is so simply assessed. In her book *Thinking in Bets*, poker champion Annie Duke (2018) contends that life is not like chess but like poker—subject to imperfect and asymmetrical information, requiring players to make the best possible decisions based on what they know about a situation at this very moment.[8] Duke warns readers against the concept of *resulting*, which is when someone focuses on the result of a choice rather than the soundness of the decision. When a player presses their luck on a hand, they are making a choice, and whether that is a good decision or a bad decision—good play/performance or bad play/performance—is not entirely up to them. A bad result can come from a good choice.

Just as even the most expert poker player cannot guarantee the outcome of a round, we as consultants cannot guarantee a grade or a product. What we can do, however, is help writers think of their process as a kind of performance and a series of choices, which includes helping them navigate between safe moves and risky ones in their writing. Instead of focusing on the grade as a score, we can help students understand writing as play by helping them examine the moments and moves they were scored on—casting us as coaches in a very literal way, who might work with students to unpack the lessons they can carry from one performance to their next. Doing so situates our work in the uncertain, turbulent, productive space between *ludus* and *paidia*, between the explicit and the implicit, the structured and the exuberant—a space where we can help students see writing as an almost athletic performance, something to be strengthened, refined, played with, and celebrated. We can't win at writing, and neither can our students, but we can all play at our best.

PLAYING WITH CADENCE

If writing is a kind of play and writing centers are a space for playing, then it stands to reason that we should understand what kinds of play are available to us and to the writers we serve. If *paidia* and *ludus* represent a play experience's relationship to rules—either diminishing them or embracing them—then we can only help writers play *better* if we can recognize these influences and make them explicit. In doing so, we can help the writers we serve to draw on the power and engagement that are uniquely concentrated in play and games.

Paidia is a powerful remedy for both boredom and despair. The paidic spirit is already embodied in a number of common writing center techniques and strategies: free writing, games, icebreakers. Boquet's (2002) opening to *Noise from the Writing Center* reminds us that when we are playful, noisy, and productive, an outsider listening in might mistake a tutoring session for a party. To this end, *paidia* is not a momentary distraction from work but an active posture.

Consider Sun Young. In her article on play in the writing center, scholar Nancy Welch (1999) relays Young's story as a student who despairs of writing but is naturally drawn to poetry. The tutor working with her sees this and leverages it, encouraging her to create her outlines, thoughts, and notes as if they were poems themselves. Welch stresses that this play "is not some detour in an otherwise one-way road towards assimilation" but instead an essential function of this student-consultant relationship, one that resists "the idea that academic writing must be an entirely joyless and uncreative activity" (61). In this way, the consultant worked alongside Sun Young, trusting a paidic instinct to find a new and unexpected path to progress.

Ludus is likewise valuable to a consultant because it illuminates how a restriction—whether a rule, challenge, or other obstacle—can actually create more engagement rather than less. This often leads us back to the metaphor of the consultant-as-coach; instead of being purely a motivating force, this coaching role is also about setting challenges to help the writer develop and improve. This can be a fraught prospect for many writing center professionals because it runs the risk of returning to the feeling of the grammar fix-it shop, with drills and rote exercises. Writing center scholarship spent decades resisting that model and breaking free of its assumptions—that students could be sent to the center to be fixed, that problems with writing were primarily sentence-level issues, that what mattered was the product, not the process. Implemented poorly or halfheartedly, ludic play can drift toward any and all of these issues, but it doesn't have to.

In examining the use of Lev Vygotsky's concept of the zone of proximal development, John Nordolf (2014) outlines two types of scaffolding used by writing consultants: cognitive and motivational. In the first case, the consultant is more directive, focused on explaining, simplifying, or connecting concepts; in the second, the consultant is nondirective, providing sympathy and support. These two forms of scaffolding help the writer perform in the zone of proximal development—the Goldilocks space between what they can accomplish on their own and what they can't accomplish, even with help. Isabelle Thompson (2009) makes a similar argument in favor of scaffolding: "In this asymmetrical relationship, the more expert tutor is expected to support and challenge the less expert student to perform at higher levels than the student could have achieved without assistance . . . In other words, tutors balance between encouraging student responsibility and ownership and guaranteeing successful student performance" (419). The benefit of *ludus,* then, is that it offers an avenue between directive and nondirective feedback. Standing somewhere in between these two extremes (or away from them entirely), the tutor-as-coach can be directive in issuing challenges or developing skills without assuming ownership or control over a student's paper.

In my own practice, I work with dozens of students every semester who struggle with thesis statements. When I approach these sessions from a ludic angle, it often begins by discussing theses and asking them to write a handful of potential statements. We move through each of them and consider what effect they may have on the paper, what inferences they make. Rather than directly evaluating one thesis to which the student is already committed, this small challenge gives us a chance to look at the strengths and weaknesses of certain choices, leaving the writer ultimately in control of the final decision.

So here is the dance: *paidia* lures us away from moribund structures, and architecturally minded *ludus* helps us build them anew. Moving between them purposefully gives the consultant a greater awareness of each session as a play space, one they can help design to guide a student toward greater performance.

THE CENTER IS THE PLAYGROUND

In *Play Anything,* Ian Bogost (2016) reflects on a trip to the mall with his daughter, unremarkable in almost any aspect except in how he notices that his daughter quietly invented a game for herself to avoid stepping on cracks. This isn't really unexpected because children are natural players, but Bogost notes that the time and space his daughter carves out to play this game—this mall, this trip—is a magic circle, one that circumscribes the situation of play to help separate it from the ordinary world. But Bogost ultimately rejects the term *magic circle,* albeit gently; instead, he likens these circumstances to playgrounds, defined by their boundaries and their contents.[9]

With this small replacement, Bogost makes his case for playgrounds not as separate from the world but as places where individuals engage fully and enthusiastically with a small subsection of it. For Bogost, play is "not an act of diversion, but the work of working a system, of interacting with the bits of logic within it. Fun is not the effect of enjoyment released by a system, but a nickname for the feeling of operating it" (114).

This spirit of wholeheartedly engaging with a space and the opportunities it provides is seen also in Boquet's (2002) call for writing centers to embrace their own liminality: "Our work is, of course, not without order, nor should we want it to be. But from whence is that order derived? If the writing center is to function as an apparatus of educational transformation, that order must develop out of chaos, not through the elimination of it. We must imagine a liminal zone where chaos and order coexist. And we would certainly do a service to ourselves, to our students, and to our institutions if we spent as much time championing the chaos of the writing center as we do championing the order" (84).

Although not explicitly named, it is hard not to see play as an essential element of this chaos, a vitalizing force often pushed to the fringes in education. This allows us to embody Nancy Welch's (2010) critical exile, embracing the center's position as a place "freed from the constraints of a predetermined curriculum and the normative force of grades" (71).

Whatever the form play takes in our centers, it is neither easy nor without pitfalls. In his own argument for play, Scott Miller (2008) makes a special note of the fact that play helps us learn how to live within paradoxical systems. He asks of us all: "How much less successful would we writing center folks be at living in our contrary, contradictory institutional site if we had not learned how to play?" (37). In learning how to play, we can better develop our instincts through the lens of *paidia* and *ludus* as a tool for understanding what kind of play is happening and how we can respond to it, focusing our attention on how that play can be changed through either childlike chaos or the orderly rigors of the chessboard.

DOING THE FUN

Writing center work is hard; writing center play may be just as hard, if not more so. For writing centers to draw strength from play requires a willingness to look, at least at times, a little bit foolish: to step back with writers and help them fiddle and bend their writing into weird shapes, even when doing so seems unproductive. To issue challenges in the spirit of *ludus*, encouraging writers to cut words, seek out weak phrases, and see each paragraph as a performance that can be tightened. Most of all, it requires us all to become more comfortable moving between these two extremes, swinging from *paidia* to *ludus* and back again in a way that helps students see these modes of engagement not as opposites but as cyclical and reciprocating forces that feed and enrich one another.

The fun we need to capitalize on (for whatever definition we want to invoke of that term) can be found in our attitude and posture toward the center-as-a-playground: "Fun is not a feeling so much as an exhaust product when an operator can treat something with dignity" (Bogost

2016, 87). Fun is in the *doing*, and so, too, is play. This is the key to understanding playfulness: that it is in the spirit of *playing with*—as in playing with toys, with friends, with words, and with writing. In helping writers find that playfulness through both freedom and challenge, we can make room for noise, for chaos, for liminality and contradictions, messy drafts and beautiful accidents.

NOTES

1. For puns and language play in the writing center, see Hall (2016). Play is also a common element throughout *Creative Approaches to Writing Center Work* (2008), especially Miller (21–49), McGlaun (115–34), and Cerbais (135–46).
2. In their chapter in this collection, Brenta Blevins and Lindsay A. Sabatino also explore *paidia* and *ludus*, likening them to the exploratory stage of writing and the drafting/editing stage, respectively. Their examination of the *paidia-ludus* continuum in writing consultations includes a more detailed look at the cues that signal each type of play.
3. Thomas "Buddy" Shay and Heather Shay explore role playing more thoroughly in their chapter in this collection, especially in conceptualizing how tutors can understand tutoring through the lenses of character and persona.
4. Role-playing exercises are also common in tutor training; see Ryan and Zimmerelli (2016, 138–48).
5. While academic writing encompasses many distinct forms and genres, I am collapsing them into this term for the sake of brevity. While it would be difficult to fully articulate academic writing in a way that is 100 percent consistent with every possible example, the underlying assumptions about a certain level of sophistication and disciplinary awareness are common in the writing students are asked to produce.
6. While *ludus* often includes a way to successfully end a game, Caillois (1961) includes open-ended hobbies as an example of ludic play, which lack clear win conditions other than those decided upon by the hobbyists themselves. Can the writers we work with *win*? It depends entirely on what they see as their end goal.
7. This is a philosophy so important it is covered on page 3 of Ryan and Zimmerelli's (2016) essential *The Bedford Guide for Writing Tutors.*
8. Duke was also a National Science Foundation fellow during her graduate studies in cognitive linguistics.
9. In their chapter in this collection, Kevin J. Rutherford and Elizabeth Saur go into more detail about the concept of the magic circle and writing centers. They draw on scholars like Castronova, Juul, and Consalvo to articulate magic circles as less separate and more porous than earlier definitions, which aligns well with Bogost's metaphor of playgrounds.

REFERENCES

Bergmann, Linda S. 2010. "The Writing Center as a Site of Engagement." In *Going Public: What Writing Programs Learn from Engagement*, edited by Shirley K. Rose and Irwin Weiser, 160–76. Logan: Utah State University Press.

Bogost, Ian. 2016. *Play Anything: The Pleasure of Limits, the Uses of Boredom, and the Secret of Games.* New York: Basic Books.

Boquet, Elizabeth. 2002. *Noise from the Writing Center.* Logan: Utah State University Press.

Busekrus, Elizabeth. 2017. "Kairotic Situations: A Spatial Rethinking of the Burkean Parlor in the Writing Center." *Praxis: A Writing Center Journal* 14, no. 2: 15–20.

Caillois, Roger. 1961. *Man, Play, and Games.* Translated by Meyer Barrash. New York: Free Press of Glencoe.

Carino, Peter. 1992. "What Do We Talk about When We Talk about Our Metaphors: A Cultural Critique of Clinic, Lab, and Center." *Writing Center Journal* 13, no. 1: 31–42.

Cerbais, Chad. "Incorporating Play and Toys into the Writing Center." In *Creative Approaches to Writing Center Work,* edited by Kevin Dvorak and Shanti Bruce, 135–46. Cresskill, NJ: Hampton.

Dixon, Dan. 2009. "Nietzsche contra Caillois: Beyond Play and Games." Paper presented at the Philosophy of Computer Games Conference, Oslo. August 1.

Duke, Annie. 2018. *Thinking in Bets: Making Smarter Decisions When You Don't Have All the Facts.* New York: Portfolio.

Feltenberger, Alaina, and Allison Carr. 2011. "Framing Versatility as a Positive: Building Institutional Validity at the University of Colorado at Boulder's Writing Center." *Praxis: A Writing Center Journal* 8, no. 2. http://www.praxisuwc.com/feltenberger-carr-82.

Hall, Amelia. 2016. "Tutor's Column, Playfulness in Discipline: How Punning Witticism is Transforming Criticism in the Writing Center." *WLN: A Journal of Writing Center Scholarship* 40, no. 9–10: 23–26.

Harris, Muriel. 1992. "The Writing Center and Tutoring in WAC Programs." In *Writing across the Curriculum: A Guide to Developing Programs,* edited by Susan H. McLeod and Margot Soven, 109–22. Newbury Park, CA: Sage.

Henricks, Thomas. 2010. "Caillois's Man, Play, and Games: An Appreciation and Evaluation." *American Journal of Play* 3, no. 2: 157–85.

Jensen, Graham H. 2013. "Making Sense of Play in Video Games: *Ludus, Paidia,* and Possibility Spaces." *Eludamos: Journal for Computer Game Culture* 7, no. 1: 69–80.

Lakoff, George, and Mark Johnson. 1980. *Metaphors We Live By.* Chicago: University of Chicago Press.

Lucero, Andres, Evangelos Karapanos, Juha Arrasvuori, and Hannu Korhonen. 2014. "Playful or Gameful? Creating Delightful User Experiences." *Interactions* 21, no. 3: 35–39.

Lunsford, Andrea. 1991. "Collaboration, Control, and the Idea of a Writing Center." *Writing Center Journal* 12, no. 1: 3–10.

McGlaun, Sandee. 2008. "Putting the 'Play' Back into Role-Playing: Tutor Training through Interactive Performance." In *Creative Approaches to Writing Center Work,* edited by Kevin Dvorak and Shanti Bruce, 115–24. Cresskill, NJ: Hampton.

Miller, Scott. 2008. "Then Everybody Jumped for Joy! (But Joy Didn't Like That, So She Left.)" In *Creative Approaches to Writing Center Work,* edited by Kevin Dvorak and Shanti Bruce, 21–46. Cresskill, NJ: Hampton.

Nordolf, John. 2014. "Vygotsky, Scaffolding, and the Role of Theory in Writing Center Work." *Writing Center Journal* 34, no. 1: 45–84.

Ryan, Leigh, and Lisa Zimmerelli. 2016. *The Bedford Guide for Writing Tutors.* 6th ed. Boston: Bedford/St. Martin's.

Sicart, Miguel. 2014. *Play Matters.* Cambridge, MA: MIT Press.

Thompson, Isabelle. 2009. "Scaffolding in the Writing Center: A Microanalysis of an Experienced Tutor's Verbal and Nonverbal Tutoring Strategies." *Written Communication* 26, no. 4: 417–53.

Welch, Nancy. 1999. "Playing with Reality: Writing Centers after the Mirror Stage." *College Composition and Communication* 51, no. 1: 51–69.

Welch, Nancy. 2010. "From Silence to Noise: The Writing Center as Critical Exile." *Writing Center Journal* 30, no. 1: 69–89.

Wilson, Nancy Effinger. 2012. "Stocking the Bodega: Towards a New Writing Center Playground." *Praxis: A Writing Center Journal* 10, no. 1. https://repositories.lib.utexas.edu/handle/2152/62149.

2

COMPLICATING GAME AND PLAY METAPHORS

The Potential for Game Heuristics in the Writing Center

Neil Baird and Christopher L. Morrow

In summer 2018, the outgoing writing center director at Western Illinois University, who had recently taken a position elsewhere, sat down for a number of training sessions and conversations with the incoming director. During this transition period, we worked through a number of issues, from protecting the writing center—given Western's financial difficulties—to strengths and weaknesses of each continuing consultant. Our conversations about continuing consultants focused particularly on Libbie, a relatively new undergraduate consultant. A late hire replacing a consultant who had recently graduated, Libbie completed her writing center training (a series of theoretical and practical readings, peer observations, and mock sessions supported by written reflection and one-on-one conversation with the director), but not without difficulty.

As a first-year student in her second semester and one of only six undergraduates in a center staffed predominantly by graduate students, Libbie lacked experience, even though she was a good writer with strong analytical and interpersonal skills. Because of her inexperience, Libbie struggled with prioritizing writing concerns, developing appropriate questions and approaches to address those concerns, and dealing with difficult students and situations. Despite these challenges, the director felt her major in education and her writing background would, over time, help her develop into an exceptional consultant. Unfortunately, continued observations by the director and peer consultants, supported by student evaluations, suggest that Libbie still struggled more than a year later. Neither staff training meetings nor director and peer mentoring were effective in helping Libbie overcome her challenges as a consultant.

Because the incoming director had a background in game studies, conversations between the two directors often explored what role game

https://doi.org/10.7330/9781646421947.c002

theory might play in the writing center. In one of these conversations, Libbie's progress once again featured prominently. Because conventional training and development practices had not been effective, could game studies provide the answer? As the two directors talked about Libbie's challenges within the context of game studies, they began to realize the potential of game heuristics to help her progress. Conventionally, game heuristics are useful to game designers in testing and refining their products. However, as we will explore in this chapter, there is an area of game heuristics that focuses on players. Like other chapters in the collection, we do touch on the designer perspective, but our chapter focuses on how player heuristics can be useful for three writing center stakeholders specifically: directors, consultants, and writers. In doing so, we draw on our administrative experiences and Libbie's experiences as a consultant as case studies to concretely ground our theoretical discussion.

"Heuristic" comes from the Greek word εὑρίσκω, which means to find or discover. Heuristics are commonly defined as practical, often systematic, approaches to discovery or problem solving. Many disciplines employ heuristics to guide their activities and the production of knowledge, such as a doctor's use of decision-making heuristics in medicine or the "gaze heuristic" employed by baseball players to catch a fly ball. Heuristics often take the form of axioms or rules of thumb, which are then used to revise a product or identify the best decision in a particular situation, such as always buying railroads in *Monopoly*. Often, the axioms themselves, once established, are primarily static; that is, selected and implemented without situational consideration. For example, hold Australia to win at *Risk* is an example of such a static heuristic.

While the gaming industry uses static heuristics to examine usability and value, game studies also offers dynamic heuristics. In dynamic heuristics, axioms are applied to a situation but adapt in response to changes in the situation. Such heuristics are deployed not by developers but rather by players who use them to negotiate the game. Although heuristics have a rich history in writing centers, as we argue below, they are primarily conceptualized as static. For example, several heuristics have been designed to aid consultants in developing effective questions, but such heuristics are presented as universal, applying to all situations. In game studies, a player's heuristics adapt as more information is learned about the game state. Dynamic heuristics help players assess their progress in a game. Any game that keeps a running score provides an understanding of how well each player is doing (positional heuristics). With this information, players develop strategies to increase their score (directional heuristics). In a more specific example, if you were playing *Settlers of Catan,* understanding your score in relation to the scores of others (positional heuristics) might prompt you to acquire more development cards for points (directional heuristics). Furthermore, through the concept of the heuristic circle, these dynamic heuristics not only change from context to context but change multiple times within the same game.

Before turning more specifically to game heuristics, we examine the history of heuristics in writing center work, then explore and outline the potential for game heuristics to assist directors, consultants, and writers. Finally, we finish with questions to guide future RAD (replicable, aggregable, data-driven) research. Ultimately, we argue that the dynamic nature of game heuristics holds the potential for complicating the "game" and "play" metaphors first suggested by Daniel T. Lochman (1986) and offers administrators, consultants, and writers new, context-specific, and innovative ways to improve their centers, their sessions, and their writing.

HEURISTICS IN WRITING CENTER STUDIES

To realize the potential of game studies heuristics, we briefly review heuristics in writing center studies. In table 2.1, we summarize writing center scholarship focused on heuristics. As the table suggests, heuristics have organized the work of three writing center stakeholders: directors, consultants, and writers. In this section, we illustrate each of these three areas with one specific example from scholarship. To problematize heuristics in writing center studies and to create a space for games heuristics, we then demonstrate how Libbie employed a heuristic featured prominently in our writing center training.

Many heuristics in writing center (WC) scholarship have been created for directors, enabling them to systematically think through issues such as assessment, relocation, and institutional change. Elizabeth Vincelette (2017), for example, offers a heuristic to help directors explore issues related to moving a writing center to a new location. Developed from her own experience relocating Old Dominion's writing center, she provides a list of questions in five areas: policy, budget, physical space, collaboration, and labor (23–24). Regarding physical space, for instance, directors are encouraged to think about how the design of the new space, from its size and furnishings to cleaning and security, impacts the following questions: "How might the allotted space for the WC impact the number of sessions? Will space expansion or

Table 2.1. Writing center heuristics

Stakeholder	Purpose for Heuristic	Scholarship
Directors	USABILITY: aids directors in evaluating usability of writing center communication	Brizee, Souza, and Driscoll (2012)
	ASSESSMENT: aids directors in matching data collection methods to assessment questions	Lerner and Kail (2014)
	INSTITUTIONAL CHANGE: aids directors in making decisions concerning how writing centers respond to institutional change (based on organization development theory)	Brady and Singh-Corcoran (2016)
	RELOCATION: aids directors in making decisions about moving writing center locations	Vincellete (2017)
Consultants	NONDIRECTIVE TALK: aids consultants in forming nondirective questions for students in four areas: establishing rapport, exploring potential, discovering strategies, and ongoing self-review	Ashton-Jones (1988)
	CONSULTING STYLE DECISION TREE: aids consultants in deciding among directive, nondirective, or collaborative approaches (in the form of a decision tree)	Henning (2005)
	DOMAINS OF EXPERTISE: aids consultants in forming questions for students using Beaufort's five domains of expertise: subject matter, writing process, rhetoric, genre, and discourse community	Savini (2011)
	OPPRESSIVE LANGUAGE: aids consultants in identifying and challenging oppressive language in writing center talk and student writing	Suhr-Sytsma and Brown (2011)
Writers	ARISTOTELIAN: shared with writers to aid invention (based on topoi)	Olson and Alton (1982)
	READING: shared with writers to aid evaluation of sources for research writing in six categories: authority, currency, relevancy, accuracy, objectivity, and appropriateness	Horning (2017)

constraint impact hours? How does the current number of tutors and scheduling affect the space? How does the space affect the number of tutors on staff? How many people could comfortably collaborate in the WC space?" (23). Her heuristic thus helps directors think through how changes to physical space impact writing center policy, budget and labor issues, and opportunities to collaborate.

A second set of heuristics has been created for consultants, aiding question formation in tutorials, decisions about particular approaches to students and situations, and even how to critique writing center talk itself. Responding to such texts as *Facing the Center* and *Writing Centers and the New Racism*, Mandy Suhr-Sytsma and Shan-Estelle Brown (2011) present a two-list heuristic, developed from focus groups with consultants, to aid consultants in promoting discussion of language, oppression, and resistance. The first list, How Language Can Perpetuate Oppression, offers ways the everyday language of both consultants and students supports oppressive practices, such as avoiding discussion of differences and misrepresenting views in sources. The second, How Tutors and Students Can Challenge Oppression through Attention to Language, lists ways of challenging oppressive language in the writing center, such as clarifying meaning together and naming "the elephant in the room" (22). They invite other centers to adapt their lists based on the provocative discussions and improved consulting fostered by the heuristic.

Finally, certain heuristics have been designed to be shared with writers within the context of tutorials to help them deal with particular rhetorical situations, such as invention and evaluating sources. In a special *WLN* issue on reading, Alice S. Horning (2017) argues that writing centers need to take the relationship between writing and reading more seriously, and she offers a heuristic consultants can share with writers to help students better evaluate texts. In addition to asserting that a writing center's staff should include expertise in reading and that consultant training needs to draw on reading scholarship, Horning offers three strategies consultants can use to help students read more effectively, one of which is a heuristic. This heuristic offers students a list of questions under six categories—authority, currency, relevancy, accuracy, objectivity, and appropriateness—to aid them in evaluating sources for research projects. Under "authority," for example, the heuristic encourages students to ask such questions as who is the author; what qualifications does the author have; and, if information is hosted on a webpage, who hosts the page (5).

In Western Illinois University's writing center, Catherine Savini's heuristic is prominent in consultant training; it's also a heuristic the outgoing director used specifically with Libbie to help her develop the ability to collaborate more effectively with students. In "An Alternative Approach to Bridging Disciplinary Divides," Savini (2011) offers a heuristic that helps consultants move beyond asking questions primarily about writing process and rhetoric. Drawing on Anne Beaufort (2007), she offers a bulleted list of questions connected to five domains of writing expertise—subject matter, writing process, rhetoric, genre, and discourse community—and proposes that the genre knowledge question "do you have a model?" is the most important question a consultant can ask to help students enact writing transfer (Savini 2011, 4).

In his observations, the outgoing director noticed that Libbie had a tendency to tell students what needed to be revised rather than collaboratively approach a revision strategy. Asking open-ended questions would have helped Libbie learn more about students' goals and thus be able to prioritize their concerns more effectively. In follow-up conversations, Libbie explained that her directive approach was a combination of feeling nervous, having difficulty leaving behind the familiar editorial role, and not knowing what questions to ask to foster collaboration. Believing that Savini's heuristic would help resolve some of these issues, the director and Libbie spent time studying Beaufort's five domains of expertise as well as the kinds of questions Savini encourages consultants to ask within each of the domains. However, in Libbie's next session, the director observed her employ questions from the heuristic, but the session never really got off the ground.

When the director talked with her afterward, Libbie expressed that she felt overwhelmed because she didn't know what to ask. That is, she now had a repertoire of questions to ask, but should she ask questions about subject matter? Or genre? Or process? What we came to realize was that Libbie's response reflected a fundamental problem with writing center heuristics: their static nature. For example, regarding her reading heuristic, Horning (2017) argues, "I find the questions associated with its six topics can effectively be applied to any kind of source material" (5). But can heuristics be applied universally? That is, can heuristics be applied wholesale from one context to another? Libbie suggests otherwise. The ever-changing context of a tutorial greatly impacts the way heuristics are employed, and Libbie lacked the ability to analyze this context to make decisions about which aspects of Savini's heuristic to adapt. Instead, we believe games studies heuristics, because of their dynamic nature, offer powerful possibilities within the context of writing center work.

HEURISTICS IN GAME STUDIES

Like writing center heuristics, heuristics in game studies target different stakeholders—in this case, game designers and players (see table 2.2). In this section, we briefly summarize designer heuristics but describe player heuristics in depth, focusing specifically on the heuristic circle and positional/directional heuristics. While we believe designer heuristics can be applied to writing centers, we believe the dynamic nature of player heuristics can significantly impact writing center stakeholders.

Designer heuristics principally serve to guide the development of the game product. Similar to writing center heuristics, these heuristics rely on static axioms to help produce an assessment. For example, one of Heather Desurvire and Charlotte Wiberg's (2009) game-play axioms

Table 2.2. Game studies heuristics

Stakeholder	*Purpose for Heuristic*	*Scholarship*
Designers	GAME USABILITY HEURISTICS (PLAY): aids designers in measuring game design across three principal categories: • GAME PLAY • COOLNESS/ENTERTAINMENT/HUMOR/EMOTIONAL IMMERSION • USABILITY • GAME MECHANICS	Desurvire and Wiberg (2009)
	INVERSE USABILITY HEURISTICS: assesses game-specific usability problems through field surveys and observations and develops inverse heuristics to assess identified problems	Pinelle, Wong, and Stach (2008)
	VALUES: aids game designers in purposefully designing values in games through a three-step process: • DISCOVERY: identify relevant values • IMPLEMENTATION: translate values into specific game elements • VERIFICATION: test and confirm that values operate as designed	Flanagan and Nissenbaum (2014)
Players	HEURISTIC CIRCLE OF GAME PLAY: aids player in navigating game through a method of exploration that develops via iterative assessments and temporary axioms	Perron (2006)
	HEURISTIC CIRCLE IN STRATEGY GAMES: aids player in developing strategic plans, identifying three levels of plans (operational, mobilized, and projected) based on three game states (immediate, inferred, and anticipated)	Dor (2014)
	POSITIONAL: aids player in determining their position within the game state with respect to game objectives	Elias, Garfield, and Gutschera (2012)
	DIRECTIONAL: aids player in determining the best strategy for reacting to a positional heuristic and improving game state position	Elias, Garfield, and Gutschera (2012)

is "the first ten minutes of play and player actions are painfully obvious and should result in immediate and positive feedback for all players" (561). Whether focused on usability or the inclusion of specific values, these static heuristics function exclusively in the design process. They help game developers identify issues early and then produce enjoyable and playable games. Once the game has been published, the heuristics are no longer necessary.

We believe there are potential applications of both usability and value-based heuristics in writing centers and call attention to their potential when highlighting future RAD research, but these heuristics are similar to current writing center heuristics, and their potential would be limited to reframing these heuristics. On the other hand, the dynamic heuristics that target players fundamentally shift the way heuristics work and offer exciting opportunities in writing centers to help directors, consultants such as Libbie, and student writers. The heuristic circle, along with positional and directional heuristics, is used not by the designer but by the player. They function not to assist in producing a better game but to assist the player in achieving their game-play goals. In this view, players, whether conscious of them or not, use heuristics to interpret the current game situation as well as successfully navigate the game. Players develop their heuristics over time to guide their game play in increasingly sophisticated ways. These dynamic heuristics not only evolve within the context of one game, they evolve over time as players return to the game. Thus these heuristics are more suited to the dynamic nature of writing centers where consultants are continually facing new challenges and situations as well as seeing the same clients over multiple sessions. (Along similar lines, Brenta Blevins and Lindsay A. Sabatino's chapter in this collection draws on the dynamic nature of game studies to develop a theory of emergent tutoring that reacts dynamically to changing forms of writing.)

Bernard Perron labels this process of assessment and response a heuristic circle. In his investigation of horror games (2006), Perron defines heuristics as a "method of exploration that proceeds by successive assessments and temporary hypothesis and that doesn't guarantee an answer or precise solution . . . The heuristic circle indeed makes explicit that the spectator must first sample the information that is provided to him (bottom-up process) to better infer the knowledge or behavior that best suits the situations (top-down process)" (64). For Perron, heuristics are tools of analysis utilized by the consumers of texts rather than the producers. Specifically discussing heuristics of analyzing film genres—what he calls the heuristic circle of a movie—he maintains that the spectator begins with the data and moves from the particular to the general in the spatial metaphor of bottom-up. This process then leads the spectator to infer the schema to apply from general to particular in the spatial metaphor of top-down.[1]

Perron goes on to address these same processes in game heuristics. Unlike films, games constantly change based on player actions. Thus, drawing on Tom Heaton's concept of the circular model of game play, Perron sees these heuristics as iterative but not discrete. Moreover, they unfold in a circle rather than a line.[2] He writes: "Compared to the spectator . . . the gamer has to get into the heuristic circle of gameplay. This magic cycle is the fun of the game" (67).[3] Thus, for Perron, gamers use top-down processes to "decide on a procedure to maximize [their] actions" but also have "to rely on the data (bottom-up process) because the game creates its own states" (68). Rather than a static set of axioms applied in only one context, these heuristics consist of constantly shifting temporary axioms.

Simon Dor (2014) significantly extends Perron's heuristic circle and the development of dynamic actions. Specifically, he argues that players (here, of real-time strategy games) utilize three levels of strategic plans: operational, mobilized, and projected. These plans are influenced by and aligned with three levels of game states perceived by the player: immediate, inferred, and anticipated. These plan levels function as short-, mid-, and long-term strategies. Operational plans effect immediate change, whereas mobilized plans consist of a sequence of actions to support the projected overall plan to achieve victory. These plans are developed in relation to game states. Immediate or current game states illustrate where a player is and what a player knows. The player must infer possible game states they are unaware of and from those inferences imagine anticipated (or future) game states. In Dor's view, all of these plans and game states are constantly evolving as operational plans create changes in the immediate game state.

Like Perron and Dor, George Skaff Elias, Richard Garfield, and K. Robert Gutschera (2012) take heuristics out of the hands of the designers and put them in the hands of the players. They define heuristics as "rules of thumb that help [players] play the game" (29) and identify two types of heuristics that players utilize. Positional heuristics help players assess the game state and determine their position (i.e., how well they are doing or if they are winning), while directional heuristics provide players with a strategy to improve their position. The authors note that heuristics are "very much dependent on the player base as well as the structure of the game itself" (32). They also note that these heuristics can themselves be evaluated according to different criteria:

- Clear vs. Muddy: How easy is it to understand and use the heuristics?
- Rich vs. Sparse: How many heuristics does the game have? Do they cover most of the situations that arise during gameplay?
- Satisfying vs. Unsatisfying: Do the players find the heuristics enjoyable to execute, or do they seem like more work than fun (highly agential, of course)?
- Powerful vs. Weak: Do the heuristics provide a great deal of help in winning, or do they just nudge the heuristics used up a bit? (32)

The authors assert, for instance, that good heuristics are ones that are clear, rich, and satisfying but not necessarily powerful. They argue that players derive enjoyment from games by "climbing the heuristics tree," or developing increasingly sophisticated heuristics from multiple plays (32). It is the development of these heuristics that generates enjoyment.

To summarize, then, consider the real-time strategy game *StarCraft II* that is Dor's object of study. In *StarCraft II*, players select one of three different factions and construct buildings and an army to destroy another player's base. Scholarship on player heuristics suggests that not all heuristics are effective. For example, heuristics can be unsatisfying, such as when a *StarCraft* player discovers that heuristics that worked well in a player-versus-environment (PVE) game aren't helpful in a player-versus-player (PVP) game. Scholarship also suggests that players bring top-down heuristics to the game: a set of axioms based on past experience, knowledge of the game, and involvement in meta-discussions with other players inside and outside the game.

Experienced gamers, however, are able to adapt these top-down heuristics through analysis of game states. In other words, directional and positional heuristics are in a constant state of conversation, requiring players to continually revise top-down heuristics with bottom-up heuristics. An experienced *StarCraft* player will begin by employing a series of top-down heuristics developed from experience with past real-time strategy games, knowledge about the games (e.g., their faction, other factions, game map), participation in discussion with other players in online forums, and even changes to the games learned by reading developer patch notes. However, once the player runs a scout through the opposing player's base to see what is being built, the player gains knowledge of the game state and begins to make inferences about projected game states that lead to the development of bottom-up heuristics. This adaption of operationalized, mobilized, and projected plans doesn't happen just once; it happens recursively as the game state becomes clearer through subsequent encounters with the opponent. Such recursive adaptation of plans has the potential to aid writing center stakeholders.

Finally, player heuristics develop over time, and this concept of "climbing the heuristics tree" is important for writing center work. Gamification applies a metaphoric overlay of a game onto a situation or activity. Rather than produce a superficial game-like interface (the purpose of which is often transparent to users), we encourage the consideration of writing and writing center sessions as games. Like games, writing a paper or getting help on one has artificially imposed constraints (i.e., rules) and clear goals (i.e., grades). Kevin J. Rutherford and Elizabeth Saur's chapter in this volume also encourages us to think of sessions as games in their consideration of how the concept of the magic circle can shape the broader context of sessions and the writing center itself. By considering writing center sessions to be game-like, we can equip consultants like Libbie not with static axioms that may or may not be applicable but with methods and procedures to assess and evolve axioms, techniques, and tactics according to the immediate, inferred, and anticipated states of the session. Furthermore, we can also help writers themselves develop these same tools as they approach their own writing tasks. In summary, then, we argue, similar to Evelyn Ashton-Jones (1988), for helping consultants and students develop "heuristic abilities" (32), and we find the dynamic nature of game heuristics particularly powerful for doing so.

PLAYERS HEURISTICS FOR WRITING CENTER STAKEHOLDERS

In this section, we theorize the implications of player heuristics for directors, consultants, and students, drawing upon the concept of the heuristic circle to demonstrate the iterative relationship between positional and directional heuristics. Ultimately, we suggest that these heuristics provide a framework for directors, consultants, and students to react to and adjust for evolving game states they encounter as administrators, tutors, and writers. In doing so, we recognize that the end result for many game heuristics is to win, which invokes a problematic metaphor in the context of higher education and writing centers; we thus want to highlight the power of game heuristics to offer solutions within dynamic contexts.

Directors

One area in which we see the potential for directors to take advantage of the heuristic circle (where users utilize a recursive process of assessment and adaptation to changing situations) as well as of positional and directional heuristics is in negotiating financial, material, and personnel support for the writing center from the university more broadly. Between fiscal years 2015 and 2019, Western Illinois University suffered a number of financial difficulties, including diminishing state appropriations, lower revenue from reduced enrollment, and the lack of a state budget for two years. These difficulties resulted in the elimination of the part-time clerical support staff in spring 2016; in 2019 they culminated in the elimination of the faculty coordinator for the writing center at Western's branch campus, reduced graduate assistantship lines, and a 50 percent reduction in undergraduate student consultant funding. In his first year, the incoming director faced cuts to faculty, graduate, and undergraduate staff.

Drawing on the heuristic circle for strategy, he developed projected plans to retain as much of his staff as possible. While the immediate game state (lack of funding and the need for university-wide cuts) was clear to the director, he was forced to infer information about the game state for upper administration. In his attempts to achieve these plans, he developed mobilized plans to try to effect the desired outcomes. At the operational level, these plans had to evolve to meet changing game state conditions. For instance, the director submitted a proposal to upper administration to create a staff position to coordinate the branch university writing center. However, this plan was based on a faulty inference of the game state that believed staff funding would be easier to procure than faculty funding. In response to this shift in game state, the director mobilized two operational plans: (1) rewrite the proposal and submit it to the provost to reverse the elimination decision and (2) rewrite the proposal into a development pitch to find private funds to support the writing center.

In terms of funding for graduate and undergraduate consultants, positional heuristics indicated that the writing center, while suffering from cuts to staffing, was also uniquely positioned to assist with an increasing online student population at Western Illinois University. Directional heuristics indicated that the best way to improve this position would be to capitalize on the opportunity to provide increased services to both online and residential students. Operational plans for resurrecting a writing fellows program, initiating a pilot program for asynchronous tutoring, and providing online training courses (some of which were already in development) were mobilized as proposals that would require more support. While these proposals have not yet resulted in increased student funding, they are clearly changing the game state as administrators, engaged in their own directional heuristics, are supporting the development of this student support and are aware of the support required.

Ultimately, static heuristics are not as effective in this situation because the game state is constantly shifting—often within the context of a single meeting. Dynamic heuristics offer directors the opportunity to constantly assess and reassess game states and make adjustments to strategic plans. Finally, the first-year director is "climbing the heuristics tree." While not initially experienced in representing the needs of the writing center to administrators, the heuristics he employs are becoming increasingly sophisticated.

Consultants

Another area in which we see potential for the heuristic circle as well as positional and directional heuristics is in developing consultants' heuristic abilities. Here, we think through what player heuristics mean for reframing our work with Libbie. We do so tentatively, with the caveat that we are theorizing here. As we outline in our final section, we're eager to learn what RAD research based on game studies heuristics might teach us about writing center work. In this section, then, we outline how we are rethinking our work with Libbie, but we can't claim with much certainty what will happen as the result of reframing her training.

As we discussed our approach to training Libbie, we came to realize that we had shared with her primarily directional heuristics. For example, after studying Savini's heuristic, Libbie gained a set of questions she could ask about subject matter, writing process, rhetoric, genre, and discourse community. Also, having read the Bedford and Longman guides to tutoring, she had a set of axioms for dealing with particular students and situations. Libbie's experience suggests that the directional heuristics she received from training were too rich, perhaps making her use of heuristics muddy. What she lacked was positional heuristics. That is, when should she ask questions about subject matter rather than genre? What if an axiom employed to deal with an aggressive student doesn't work? Having considered the implications of player heuristics, we see a way forward in developing her heuristic abilities by drawing on the heuristic circle in four ways: (1) evaluating the top-down heuristics she brings to a session, (2) teaching her how to assess the immediate state of the session, (3) helping her make effective inferences about the anticipated state of the session, and (4) assisting her in mobilizing bottom-up heuristics in response to the changing state of a consulting session. In addition, Elliott Freeman's chapter in this volume—in particular, the exploration of how the concepts of *paidia* and *ludus* can be dynamically marshaled within a session depending on the situation by either introducing, enforcing, renegotiating, or removing certain rules and challenges—may offer an additional avenue to help Libbie in this situation.

Our first goal, then, is to identify and evaluate the top-down heuristics Libbie brings to a session. That is, what heuristics does Libbie bring to a session, and to what extent do these heuristics create the foundation for an effective session? Elias, Garfield, and Gutschera (2012) call the heuristics used by beginners "zero-level heuristics" and note that they are incredibly important: "Players who first learn the game need to have some idea of what they are trying to do and how they might go about it" (35). If they don't have effective zero-level heuristics, they are likely to become frustrated with the game and simply quit. A coach or team manager can help facilitate the development of zero-level heuristics, but the authors are quick to point out that too many "transmitted heuristics . . .

can create a burden on memory" (32–33). Also, other players can facilitate the development of zero-level heuristics. They note that heuristics differ by player base (35). That is, the heuristics that feature prominently in a particular board game with one group may differ substantially from those of another group. In addition, as a player base gains expertise in a game over time, the heuristics will evolve to follow suit.

What this discussion means is that learning the top-down heuristics Libbie employs is more complex than just working with Libbie. Certainly, we need to learn from Libbie about the axioms or rules of thumb she uses to initially approach a consulting session to help her build new heuristics; however, the writing center staff, as the player base, is also facilitating her development of zero-level heuristics, so we need to consider the heuristics circulating among the staff:

- What heuristics for both consulting and writing are prominent among consultants in the writing center?
- Where do these heuristics circulate in the context of the writing center, and how?
- How do heuristics differ between writing center staff members?

Once we've identified and evaluated Libbie's top-down approach, we need to learn how she assesses the immediate state of the session and how she makes inferences about future states. After reading about heuristics in game studies, we believe this is key. It's not enough to approach a session with effective zero-level, top-down heuristics. These heuristics need to adapt as more of the session's game state is learned. That is, we need to move Libbie to the heuristic circle model. When observing Libbie, this means paying attention to how she perceives the immediate state of the student, the writing, and the session and asking her questions about her perceptions afterward:

- How did you initially perceive the session, and why?
- How did your perceptions change across the session, and why?
- To what extent did your understanding of the immediate state of the session accurately reflect the session?

Helping Libbie move toward the heuristic circle model also means learning more about how she made the state of the session clearer and what inferences she made based on that knowledge:

- What strategies did you use to learn more about the writing?
- How did you learn more about the student and the particular behaviors present in the session?
- How did you proceed based on what you learned, and why?

Once we learn about how Libbie perceived the immediate state of the session, the strategies she used to learn more about the state, and how she made inferences based on that knowledge, we will be in a position to assist her in mobilizing bottom-up heuristics in response to the changing state of a consulting session. We can do so by asking her questions about how she adapted the axioms or rules of thumb prominent in her top-down approach:

- When you learned more about the draft the student brought to the session, I noticed that you asked questions about writing process. How did you adapt Savini's questions?
- Why mobilize a writing process approach over genre or discourse community?
- When you learned more about why this student was unresponsive, how did you adapt what you know about working with unresponsive students from our tutoring guides?
- How did you continue to adapt these rules of thumb as the session continued?

We believe that having this series of discussions will help develop the positional heuristics Libbie's directional heuristics so desperately need.

Students

Writing centers see many student writers with a limited understanding of their own writing and even less understanding of how to improve. For these students, writing becomes a source of frustration and anxiety, much of which spills out during writing consultations. We believe the dynamic nature of player heuristics may not only help student writers improve but may perhaps provide an opportunity to bring enjoyment through "climbing the heuristics tree." That is, similar to Gary A. Olson and John Alton (1982) and Horning (2017), who designed heuristics for consultants to share with students, we believe the heuristic circle can help develop writers' heuristic abilities.

Our plan for reframing our work with Libbie also suggests how consultants might subtly shift talk about writing. Consultants can begin by helping students identify the top-down heuristics they bring to rhetorical situations by uncovering the axioms or rules of thumb they commonly employ when they write. Once these heuristics are made visible, they can help students evaluate the effectiveness of these heuristics. More important, consultants need to help students understand the need to adapt these top-down heuristics with bottom-up heuristics by learning how students come to perceive the immediate state of the rhetorical situation and make inferences about its future state from these perceptions. As consultants teach students how to learn more about the immediate game state of their writing and make inferences based on that information, they will be able to help students mobilize bottom-up heuristics in response to this anticipated state. In this way, consultants can help students assess a draft through positional heuristics and develop a revision plan through directional heuristics, and this heuristic thinking may transfer across contexts (Devet 2015) while creating enjoyment for writing.

FUTURE RAD RESEARCH

In this chapter, we theorize how game studies heuristics might impact writing work. Here, we suggest four areas of replicable, aggregable, and data-driven research (Haswell 2005) for scholars interested in pursuing the implications of game heuristics in the context of writing centers. We begin with two areas suggested by player heuristics and end with two areas suggested by designer heuristics.

Player Heuristics

1. **The Heuristic Circle**: Game heuristics highlight the interplay of two types of knowledge: positional knowledge and directional knowledge. Successful gamers are able to employ effective directional heuristics in response to knowledge derived from positional heuristics. They are able to develop more sophisticated positional and directional heuristics in response to failure as game difficulty increases. They are also able to transfer positional and directional heuristics across similar game genres and adapt them across different games. We call on scholars to explore how these spatial metaphors can help us better understand consulting and writing expertise. How do novice and experienced consultants come to understand the immediate, inferred, and anticipated game state of the session? How do they form operational, mobilized, and projected plans based on positional insight? How do consultants develop the agility to respond as strategic plans impact game states and game states impact strategic plans? In what ways can consultants develop "heuristic circle abilities" in students?
2. **Evaluating Heuristics**: Heuristics are ultimately ways of seeing, and the choice of a heuristic entails adopting partial and incomplete ways of seeing. How does Elias, Garfield, and Gutschera's framework (e.g., clear versus muddy) and others like it help us identify and evaluate positional and directional heuristics in the writing center? What other ways of evaluating heuristics might be revealed? Ultimately, how might these frameworks critique the ways we have employed heuristics as a field?

Designer Heuristics

1. **Usability Axioms**: Since the publication of Lochman's article (1986), usability heuristics (Desurvire and Wiberg 2009; Pinelle, Wong, and Stach 2008) have created a number of axioms designed to promote enjoyment in games. Like Lochman, we see "game" and "play" as fruitful metaphors for understanding writing center work; however, we call on scholars to examine how game studies axioms complicate game and play. For example, what does Desurvire and Wiberg's (2009) PLAY framework, which offers fifty axioms, suggest about successful play and games, and what might they mean for understanding writing center work?
2. **Values at Play**: Mary Flanagan and Helen Nissenbaum (2014) highlight fifteen elements that make up a game's semantic architecture: narrative premise and goals, characters, actions in the game, player choice, rules for interaction with other players, rules for interacting with the environment, point of view, hardware, interface, game engine, context of play, rewards, strategy, game maps, and aesthetics (33–34). Each of these elements works in combination to transmit values to players. As above, we ask scholars to consider what these metaphors, which complicate game and play, mean for writing center work. More important, how do these elements considered together offer a more comprehensive understanding of values at play in writing centers? The ways writing center scholars have employed heuristics as a field?

While heuristics have a rich history in both writing center and game studies, we feel that the dynamic nature of game studies heuristics—specifically the heuristic circle along with positional and directional heuristics—offers a new way of conceiving heuristics within the context of writing centers, and we are eager to learn more about how they can contribute to the development of heuristic abilities in consultants. By complicating metaphors of play and framing our work as administrators, consultants, and writers using such concepts from game studies, we are afforded new opportunities for effectively positioning writing centers within institutional contexts and equipping consultants and students with the tools to become better consultants and writers.

NOTES

1. For more information on the cognitive psychology and film studies behind the concept of the heuristic circle, see the work of Bartlett (1977), Neisser (1976), and Bordwell (1985).
2. According to this model, players first gather information on the game state, then they take actions that create new game states. These new game states are then communicated to the player, and the player, in turn, takes additional actions.
3. Perron here also alludes to Johan Huizinga's (1950) famous concept of the magic circle, which is the separate and somewhat protected space of play.

REFERENCES

Ashton-Jones, Evelyn. 1988. "Asking the Right Questions: A Heuristic for Tutors." *Writing Center Journal* 9, no. 1: 29–36.

Bartlett, Frederic C. 1977. *Remembering: A Study in Experimental and Social Psychology.* Cambridge, UK: Cambridge University Press.

Beaufort, Anne. 2007. *College Writing and Beyond: A New Framework for University Writing Instruction.* Logan: Utah State University Press.

Bordwell, David. 1985. *Narration in the Fiction Film.* Madison: University of Wisconsin Press.

Brady, Laura, and Natalie Singh-Corcoran. 2016. "A Space for Change: Writing Center Partnerships to Support Graduate Writing." *WLN: A Journal of Writing Center Scholarship* 40, no. 5–6: 2–9.

Brizee, Allen, Morgan Souza, and Dana Lynn Driscoll. 2012. "Writing Centers and Students with Disabilities: The User-Centered Approach, Participatory Design, and Empirical Research as Collaborative Methodologies." *Computers and Composition* 29, no. 4: 341–66.

Desurvire, Heather, and Charlotte Wiberg. 2009. "Game Usability Heuristics (PLAY) for Evaluating and Designing Better Games: The Next Iteration." *OCSC 2009 Lecture Notes in Computer Science* 5621: 557–66. Springer, Berlin, Heidelberg.

Devet, Bonnie. 2015. "The Writing Center and Transfer of Learning: A Primer for Directors." *Writing Center Journal* 35, no. 1: 119–51.

Dor, Simon. 2014. "The Heuristic Circle of Real-Time Strategy Process: A StarCraft, Brood War Case Study." *Game Studies* 14, no. 1. http://gamestudies.org/1401/articles/dor.

Elias, George Skaff, Richard Garfield, and K. Robert Gutschera. 2012. *Characteristics of Games.* Cambridge, MA: MIT Press.

Flanagan, Mary, and Helen Nissenbaum. 2014. *Values at Play in Digital Games.* Cambridge, MA: MIT Press.

Haswell, Richard. 2005. "NCTE/CCCC's Recent War on Scholarship." *Written Communication* 22, no. 2: 198–223.

Henning, Teresa. 2005. "The Tutoring Style Decision Tree: A Useful Heuristic for Tutors." *WLN: A Journal of Writing Center Scholarship* 30, no. 1: 6–8.

Horning, Alice S. 2017. "Reading: Securing Its Place in the Writing Center." *WLN: A Journal of Writing Center Scholarship* 41, no. 7–8: 2–8.

Huizinga, Johan. 1950. *Homo Ludens: A Study of the Play-Element in Culture.* Boston: Beacon.

Lerner, Neal, and Harvey Kail. 2014. "A Heuristic for Writing Center Assessment." Presentation given at the Summer Institute for Writing Center Directors, July 12–19, 2014, Clark University, Worcester, MA.

Lochman, Daniel T. 1986. "Play and Games: Implications for the Writing Center." *Writing Center Journal* 7, no. 1: 11–18.

Neisser, Ulric. 1976. *Cognition and Reality: Principles and Implications of Cognitive Psychology.* New York: W. H. Freeman.

Olson, Gary A., and John Alton. 1982. "Heuristics: Out of the Pulpit and into the Writing Center." *Writing Center Journal* 2, no. 1: 48–56.

Perron, Bernard. 2006. "The Heuristic Circle of Gameplay: The Case of Survival Horror." In *Conference Proceedings medi@terra–Gaming Realities: A Challenge for Digital Culture,* edited by Manthos Santorineos, 62–69. Athens, Greece: Fournos Center for Digital Culture.

Pinelle, David, Nelson Wong, and Tadeusz Stach. 2008. "Heuristics Evaluation for Games: Usability Principles for Video Game Design." In *Proceedings of the Twenty-Sixth Annual SIGCHI Conference on Human Factors in Computing Systems,* 1453–62. New York: Association for Computing Machinery.

Savini, Catherine. 2011. "An Alternative Approach to Bridging Disciplinary Divides." *WLN: A Journal of Writing Center Scholarship* 35, no. 7–9: 2–5.

Suhr-Sytsma, Mandy, and Shan-Estelle Brown. 2011. "Theory in/to Practice: Addressing the Everyday Language of Oppression in the Writing Center." *Writing Center Journal* 31, no. 2: 13–49.

Vincellete, Elizabeth. 2017. "From the Margin to the Middle: A Heuristic for Planning Writing Center Relocation." *WLN: A Journal of Writing Center Scholarship* 41, no. 5–6: 22–25.

3

THE BINDING OF PROCESS

Bringing Composition, Writing Centers, and Games Together

Jason Custer

Game studies, composition studies, and writing center studies have seen increasingly overlapping interests since the mid-2000s. Starting with landmark texts like Ian Bogost's (2007) *Persuasive Games*, Gunther Kress's (2010) *Multimodality*, and David M. Sheridan and James A. Inman's (2010) *Multiliteracy Centers*, these branches of scholarship have paid greater attention to concepts such as digital and multimodal rhetoric. As Stephanie Vie and Holly Ryan demonstrate in this collection's introduction, the rising interest in digital and multimodal texts across composition studies makes it increasingly common—especially given their inherently digital and multimodal nature—for computer games to garner attention from writing center studies. What we see discussed less often is the incredible overlap of one of the most vital concepts connecting computer games, digital composition, and their possibilities for supporting the increasingly digital work done in writing centers—process.

Using process as a connecting thread, this chapter demonstrates how process theory in composition studies influenced procedurality in game studies and why that matters for writing center practitioners. Given the lengthy and trepidatious relationship between writing centers and computer technology (Brown 1990; Carino 1998; Palmquist 2003; Summerfield 1988), identifying the deep connections between writing process theory and procedurality in digital artifacts recontextualizes computers and computer technology in writing centers as a logical step forward. In short, processes bring digital texts and writing center pedagogy together, and this synthesis points to a way writing center studies can remain a vital branch of composition studies. Process theory is a critical part of the history of both composition studies and writing center studies; writing center scholars' and practitioners' unique foundation atop a bedrock of process theory places them in a prime position

https://doi.org/10.7330/9781646421947.c003

to continue doing what they do best—help students build awareness and skills with a variety of processes.

In this chapter, I establish process as a conceptual keystone for understanding how game studies, composition studies, and writing center studies intersect. This chapter begins by highlighting how process was instrumental to the development of composition studies as a field. Next, it demonstrates how the process movement in composition studies led to a revolution in writing center theory and pedagogy. Finally, it connects those developments to the massively influential scholarship of Janet H. Murray (1997) and Bogost that established process as a key concept in modern game studies. Seeing process across these fields presents an exigence for writing center practitioners and pedagogy to consider how focusing on concepts like play and process may help students become better writers. This chapter argues that processes—whether writing or computational—are vital to understand the history and modern study of the place where composition studies, computer games, and future avenues for writing centers intersect.

THE CELLAR—A BRIEF HISTORY OF WRITING PROCESS THEORY IN COMPOSITION

While the 1963 Dartmouth conference serves as a convenient origin story for composition studies, the development of process theory may be the most influential movement for composition studies as an academic field (Faigley 1986, 527). The early stages of the process movement stretch back to the Harvard Committee on Composition's 1892 address, which argued that the products of writing—and not the process by which they were made—were worthy of our attention (Gere 1985, 113). The first traces of the process movement gaining traction in composition studies came in the 1920s with the rise of expressivist approaches to composition and with them the development of process theory (Berlin 1987, 75). The shift to focusing on writing process is attributed largely to Raymond Weaver's 1919 text, in which he laments that the "process by which successful writers have brought their work to its final form has not been the interest of the pedagogue" (quoted in Berlin 1987, 75), rendering this common refrain in modern composition studies more than a century old.

D. Gordon Rohman and Albert Wlecke (1964) continue to build process into the foundation of composition studies in their book *Pre-Writing: The Construction and Applications of Models for Concept Formation in Writing*. They break the writing process down into three basic parts:

pre-writing, writing, and rewriting. Their linear three-step process established the language the process movement would later use (Berlin 1987, 146) and became a wildly popular representation of the writing process (Faigley et al. 1993, 5). Despite the rigid linearity of their writing model, it is important to note that their scholarship also links composition and processes to play from the onset of the process movement. Their work on the writing process also proposes that "play is not merely childish; it is one of the archetypal forms of human activity" (quoted in Nash 1984, 193). Typically, play is seen as a crucial part of the writing process itself, where writers get to play with their ideas—on their own or in a writing center (Bain 2006; Flower and Hayes 1981; Freedman 1982; Nash 1984). The proposed linearity of the writing process in Rohman and Wlecke's model, however, led to the explosion of scholarship that defined both process theory and composition studies.

Immediately after *Pre-Writing* was published, Janet Emig (1971) argued against Rohman and Wlecke's linear model of the writing process, and the research she performed in the wake of this debate proved massively influential (Faigley et al. 1993, 5). Emig's *The Composing Processes of Twelfth Graders* in particular had massive ripple effects in composition studies; in the years that followed, several foundational pieces of process theory were published (Braddock 1974; Flower and Hayes 1977; Murray 1972; Shaughnessy 1977). In fact, Flower and Hayes (1977) employed language similar to that of Rohman and Wlecke by defining stages of the writing process such as brainstorming as "a form of creative, goal-directed play" (454). Dozens more articles and books followed, placing process theory at the forefront of composition studies—all of which point to the 1970s and 1980s as the apex of process theory's influence on composition studies.

A mere decade later, composition scholarship made massive strides toward what we now refer to as post-process theory—or the "social turn" in composition, as it is commonly called. While post-process theory did not completely eradicate discussions about writing processes from composition pedagogy, it did signal a move away from the early roots of composition studies, where writing itself and composition pedagogy were predominantly seen as writer- and process-focused practices. This shift did not spell the end of process influencing composition theory or pedagogies, however (Tobin 1994, 9). In writing center studies, where the one-on-one tutorial model focused on individual writers and tutors served as a vital part of a tutee's writing process, process theory gained significant traction. In the following section, I introduce some early writing center scholarship, illustrate how writing centers took up the mantle from early composition studies by focusing on process as a concept, demonstrate that writing center scholarship built firmly on the heights of composition's process movement in the 1970s and 1980s, and finally, discuss how a focus on process and play in writing center scholarship brings writing centers and game studies together.

THE CAVES—PROCESS THEORY AND THE DAWN OF MODERN WRITING CENTER STUDIES

Lil Brannon and Stephen North (1980, 1) cement the connections between writing centers and the process movement in the introduction to the inaugural issue of *Writing Center Journal*; they argue for the writing center as "the absolute frontier of our discipline" based on its role in "the two 'seminal ideas' that define composition for a discipline: 'the student-centered curriculum' and 'concern for composing as a process.'" Throughout the 1980s, the influence of the process movement on composition studies and its connections to writing centers flourished. North (1984) began publishing on tutor education, process, and how to position tutors between tutees and their writing processes shortly after the publication of two more process-focused guidebooks for writing center practitioners (Harris 1982; Steward and Croft 1982).

In their account of the "phenomenal growth" of writing centers in the 1970s and 1980s, Joyce S. Steward and Mary K. Croft (1982) argue that "the lab movement has gone hand in hand with the emphasis upon teaching writing as a process and with a growing recognition that the writing process differs for different individuals" (3). In fact, Steward and Croft attribute a great deal of the success of writing centers in the 1980s to the process movement and writing centers' unique ability to embody process pedagogy in their practices. They advocate for writing processes as a vital part of everything in writing centers, from the philosophy of the spaces themselves to how tutors should be educated (6, 32). Similarly, Muriel Harris's (1982) *Tutoring Writing* has no shortage of emphasis on process, with the importance of the writing process in the writing center as a central theme of its chapters (Almasy 1982; Brostoff 1982; Freedman 1982). All of this came before what was perhaps the most influential piece of writing center scholarship to date—North's (1984) "The Idea of a Writing Center," which remains a foundational piece of writing center theory and scholarship.

Considered the most important and most quoted piece in writing center studies (Murphy and Law 2000, 65), North's "Idea" outlined the basic tenets of how writing centers function and, by extension, defined writing center practices and theory for decades to come. The influence of North's piece was such that for a time, 80 percent of the issues of the *Writing Center Journal* featured at least one article that referenced it (Boquet and Lerner 2008, 176). For North (1984), the "new writing center" marked the dawn of a bright new era (438), and his article ushered in a new wave of scholarship and pedagogy. North constructs the writing center as a space where writers improve their writing processes through dialectic exchanges with tutors, which he sees as the primary concern of the center (438). The best-known part of North's text remains his famous statement about what writing centers do: "in axiom form it goes like this: Our job is to produce better writers, not better writing" (438). This line can be seen as the culmination and embodiment of the process movement itself in composition studies (Howard 2006, 5). Given the massive weight and influence North's "Idea" had on day-to-day writing center practices, it is easy to overlook just how prominent process theory is in his text and thus neglect how much it seeped into the theory and practice of hundreds of writing centers for the better part of the last fifty years.

Just like the newfound focus on composing processes, play also found its footing as an important part of writing center scholarship. Specifically, Aviva Freedman (1982) highlights the potential in working with students during the exploration stage of their composing process, since they may not have "learned the pleasure of playing with ideas and bringing their intellectual experience to bear on their lives" (8). Play also became a critical concept in writing centers as practitioners frequently recommended that tutor education introduce a role-playing element to prepare tutors for working with real clients (Bannister-Wills 1984; Garrett 1982; Nash 1984; Posey 1986). Simply put, play, process, and writing center scholarship connected in the 1980s and would go on to influence Janet Murray's scholarship and, by association, future scholarship in game studies. Recent scholarship from Murray (1997) and Ian Bogost (2006, 2007, 2008) establishes how process theory moves into digital artifacts; how it led to concepts such as procedurality, procedural rhetoric, and procedural literacy; and how those terms tie back into composition and writing center studies. As I demonstrate in the following section, the connections among process, play, and writing that started in composition studies and permeated writing center theory also directly shaped highly influential game studies scholars, including Murray and Bogost.

THE DEPTHS—PROCESSES AND PROCEDURALITY IN MURRAY AND BOGOST

Up to this point, I have outlined the history of writing centers and composition's process pedagogy. In this history, we begin to see that writing centers are possibility spaces where processes are explored, whether by playing with ideas or gaining experience with—and a better understanding of—an individual's writing processes. The clients who come to writing centers and work with tutors become more aware of their own writing processes, and in several ways that may help them become better writers. Described thus, everyday writing center practice ties directly into key concepts in game studies such as procedurality and procedural literacy. As I discuss later in this section, the interactions between tutors and students in writing centers embody concepts like procedural literacy remarkably well and highlight how writing centers have immense potential to learn from and model key concepts in modern game studies. As in the preceding sections, in this section I focus primarily on providing a brief overview that highlights how conceptually intertwined process is with composition and writing centers—particularly how this impacts computer game studies—before moving on to a more complete discussion of the implications of these connections in the final section. The work of Janet Murray is a fitting place to begin this discussion.

Murray first published *Hamlet on the Holodeck* in 1997, which went on to inspire some of the most influential modern game studies scholarship focused on process. One of Murray's (1997) major contributions to studying digital texts—and game studies in particular—is her argument that digital artifacts are procedural (e.g., composed of executable rules), participatory (e.g., welcoming human interaction and manipulation of the represented world), encyclopedic (e.g., containing a high capacity of information in multiple media formats), and spatial (e.g., users can navigate it as an information repository, a virtual place, or both). *Hamlet on the Holodeck* also features one of the first formal acknowledgments that the process movement in composition studies influenced the popularization of procedurality as a concept in game studies.

Rather than simply an inference on my part, Murray (1997) specifically mentions that her theory of digital artifacts in *Hamlet on the Holodeck* was built on composition studies (5). She notes that she was "particularly influenced" as an early scholar and teacher by "Peter Elbow and Linda Flower, both of whom pioneered the teaching of writing as a process-centered, rather than product-centered, activity" (285). I want to emphasize the importance of this particular footnote—Murray, the author of a foundational text for the study of digital artifacts, explicitly ties the process movement in composition studies and writing centers to digital studies. While the work of connecting writing centers and game studies in this collection or chapter may feel cutting-edge, Murray started building on and identifying the importance of processes and their connections to writing, tutoring, and gaming almost twenty-five years ago.

Murray deserves her place in this conversation by providing a point of origin for procedurality's role in the modern study of games and rhetoric; Bogost expands on her assertion that computer games, as expressive digital artifacts, hinge on processes as one of their essential properties. In the introduction to his 2006 book *Unit Operations*, he argues that "any medium—poetic, literary, cinematic, computational—can be read as a configurative system, an arrangement of discrete, interlocking units of expressive meaning. I call these general instances of procedural expression unit operations" (ix). Here, Bogost—as did Murray in *Hamlet on the Holodeck*—asserts that computer games are worthy of serious critical examination by arguing that games, like other media, contain a series of interlocking units of expression. In *Persuasive Games,* Bogost built further on Murray's discussion of procedurality with his term *procedural rhetoric.* In other words, Bogost's watershed scholarly contribution that brought rhetoric and computer games together owes a great deal to Murray—and, by extension, to the process movement in composition and writing center studies.

Bogost (2008) defines procedural rhetoric as "the practice of using processes persuasively, just as verbal rhetoric is the practice of using oratory persuasively and visual rhetoric is the practice of using images persuasively . . . Procedural rhetoric is a general name for the practice of authoring arguments through processes" (125). Bogost explains how this works with examples such as *Animal Crossing.* As a series, *Animal Crossing* presents itself as a friendly, slice-of-life simulator in which you interact with animals, do simple maintenance to the landscape, and decorate your home, among other activities. When we adopt procedural rhetoric as a lens through which to see the mechanics present in games, however, it becomes easier to see the first entry in the series as "a game about long-term debt. It is a game about the repetition of mundane work necessary to support contemporary material property ideals" (119). When looked at through this lens, the cute exterior of a game like *Animal Crossing* gives way to a debt simulator in which the player makes their animal landlord increasingly wealthy through a cycle of working to paying off a mortgage, acquires more possessions as they work, and enters a new mortgage to expand their house to accommodate those possessions. As Bogost notes, "*Animal Crossing* creates a representation of everyday life in which labor and debt are a part" (122–23), which is likely to feel familiar to anyone living in a predominantly capitalist society.

While Bogost (2008) makes clear his "preference for videogames" when applying procedural rhetoric as an analytical lens, he encourages his readers to "see procedural rhetoric as a domain much broader than that of videogames, encompassing any medium—computational or not—that accomplishes its inscription via processes" (46). Bogost further states that "procedural rhetoric always remains open to reconsideration, objection, or expansion, whether through further procedural models or normal written and spoken discourse" (143), which further brings procedurality, writing, and computer games together. So, while Murray explicitly connects the process movement in composition studies to digital artifacts, Bogost later built directly on Murray's work to define procedural rhetoric. It merits mentioning that most discussions of Bogost's work center on procedural rhetoric, even though it is a subsection of a larger concept: procedural literacy.

Procedural literacy presents a clear path to utilizing procedural rhetoric and play in spaces such as writing centers by welcoming us to consider the implications for the way processes work in games and a variety of educational contexts. While procedural literacy predates Murray (Mateas 2005; Sheil 1980), Bogost (2007) defines procedural literacy as "the ability to read and write procedural rhetorics—to craft and understand arguments mounted through unit operations represented in code" (258). He also extends this discussion beyond coding: "Indeed, any activity that encourages active assembly of basic building blocks according to particular logics contributes to procedural literacy" (257). Whereas Murray's scholarship establishes the role of process theory in connecting composition's history to analyzing digital artifacts, Bogost's discussion of procedural literacy bridges the gap between game studies and writing centers by demonstrating that creating meaning through processes is a core part of procedural literacy and thus welcomes us to consider how writing itself fits into this discussion.

With this in mind, perhaps the most important connection among writing centers, procedurality, and computer games is made through Bogost's argument that we gain procedural literacy by engaging in play. As noted previously, writing center scholarship has a rich history of encouraging us to see play as part of the writing process (Bannister-Wills 1984; Flower and Hayes 1977; Freedman 1982; Garrett 1982; Nash 1984;

Posey 1986; Rohman and Wlecke 1964). The present collection successfully brings this discussion closer to play in the context of computer games. Jamie Henthorn's chapter in this collection further elucidates what writing center scholars have argued for decades—writing centers are excellent examples of places where play and writing processes converge. Writing centers have positioned themselves in both theory and practice as places where students can learn to experiment with thoughts and ideas—all while engaging in vital parts of their writing processes—by having conversations with tutors. Brenta Blevins and Lindsay A. Sabatino's chapter in this collection further establishes that writing centers are a locus for play, and writing centers excel at the kind of play Bogost offers in his description of procedural literacy.

Bogost sees play as a way to understand processes and argues that we become procedurally literate through play and engaging with processes; because this is not exclusive to computer games, it presents an opening for writing center practitioners to embrace writing center spaces as opportunities to model and foster procedural literacy in students by encouraging play and focusing on their writing processes. Bogost's explanation of procedural literacy is strikingly similar to the process movement in composition studies—writers gain literacy and improve as writers by *doing* writing. The idea that writers develop their writing processes by engaging in them—in much the same way computer game players develop procedural literacy by playing computer games—is a thread that runs through a substantial portion of the writing center scholarship discussed throughout this chapter. If we think about procedural literacy outside of games—as Bogost explicitly suggests we do—the writing process movement argues that writers become procedurally literate by becoming aware of their writing process, engaging in writing processes more often, and playing with words. Writing centers excel at all these tasks.

THE CATHEDRAL—APPLICATIONS AND IMPLICATIONS

A look across the history of process in composition studies, writing center studies, and game studies demonstrates just how closely connected and recursive these three areas of study are. Starting with discussions of writing process in 1892 and then returning to the present day with discussions of procedural literacy and digital artifacts, process stands out as an unexpected link among composition studies, writing centers, and computer games. Seeing these connections does more than just highlight an interesting coincidence for writing center practitioners. First, it invites us to move beyond debates about whether computers belong in the writing center, as the connections between them have been established across writing studies for decades. Second, it encourages us to remember how important writing centers are to our clients' writing processes. Finally, seeing process as a pivotal part of writing center practices presents an opportunity for writing centers to continue to be a cutting-edge part of composition studies and campus culture.

One of the most important results of embracing process as a linchpin for composition studies, writing centers, and game studies is establishing that it is neither new nor abnormal for writing centers to work with digital texts. Blevins and Sabatino's chapter in this collection discusses cutting-edge practices such as integrating augmented reality into writing centers, yet writing centers have wrestled with whether to welcome computer technology into their spaces and practices for more than thirty years (Holmes 1985, 13). Again, as the introduction to this collection demonstrates, composition studies has shown increased interest in digital and multimodal composition in pedagogy and theory and, given the overlap between the interests of composition studies and writing centers throughout their histories, it follows that writing centers do the same. However, a look across the histories of process as a concept in composition studies, writing center studies, and game studies provides a new frame of reference for reconsidering the debate, concerns, and false starts noted by Judith Summerfield (1988) and Mike Palmquist (2003) regarding incorporating computers into writing centers. This is not to say that writing centers will, should, or should not become spaces where students actively play computer games during sessions but rather that seeing processes as a natural part of both computers and writing centers may help smooth over lingering concerns about whether providing feedback on digital texts belongs in the periphery of a writing center.

For instance, while Alan Brown (1990) and Peter Carino (1998) cast skepticism onto the idea of incorporating computer technology into writing centers, consideration of process as a key concept connects digital artifacts to the pedagogy and practice of writing centers and demonstrates how closely linked they already are. Not only does their reliance on process align them, but both potentially have a great deal to learn from one another. First, applying a process-centric pedagogy to digital and multimodal composing—as was done routinely throughout the heyday of writing center scholarship focused primarily on tutoring alphabetic writing—helps clients think more carefully about how best to design new texts. In addition, it presents an important avenue for writing centers to apply their enduring strengths to new contexts. We see this happening already with the push toward reimagining writing centers as multiliteracy centers and digital studios, but even spaces that continue to identify as writing centers remain an excellent resource for clients creating texts across media to seek a practice audience for their work. No matter the medium, genre, or audience of a text, writing centers have a long, proud history of serving as a capable audience for any text a client brings in, and the long-standing expertise with key tenets of composition—and process itself—means that writing centers of all kinds can offer invaluable support for the influx of digital and multimodal texts and pedagogies.

Furthermore, a look across composition studies, writing center studies, and game studies while focusing on process highlights the importance and possibilities a process-centric approach to the writing center can bring to its daily practices and how central processes are to that context. While some may take this as far as Evelyn Posey (1986) once did and suggest that tutors conduct sessions with a handout filled with questions specifically about processes at their side, the fact remains that any writer who enters the writing center is coming to that resource as a vital part of their individual writing or composing process. With this in mind, writing center practitioners need to consider what role they serve as part of every incoming client's process and how to best respond to each client's writing practices individually. Returning to Bogost, writing centers are in a unique position to facilitate procedural literacy in their clients and guide them not only toward a better understanding of their own processes of writing but also toward thinking about how the writing center can continue to serve as a productive part of that process. Of course, this requires that writing center workers educate and mentor tutors with this in mind, but given how important process is to the lineage of the writing center, encouraging discussions about process and making clients more aware of their own process—and engaging them to play with words and ideas throughout it—is something many writing center practitioners are likely already doing, to varying degrees.

An approach to composition studies, writing center studies, and game studies that sees process as a binding force also allows us to revisit the work of Brannon (1980) and North (1984) in a different light. In the inaugural issue of the *Writing Center Journal*, Brannon and North (1980) argued for the writing center and its focus on process as "the absolute frontier of our discipline" (1), but their argument limited the scope of process to being largely alphabetic in nature. A look at process across composition and computer games, however, gives us another way to see how writing centers can continue serving as cutting-edge, interdisciplinary places for learning through many different kinds of play. Many of the questions writing center practitioners and tutors are already asking remain equally important for designers of new media, digital, multimodal, interactive, and game-based texts. These questions may include the following: Who is your audience? What are your goals as a composer? What medium of delivery makes the most sense for your message? How does this text fit into/push back against the genre conventions you're building on? While writing center tutors in the 1980s might have asked, What stage of the writing process would you like help with today, a more prudent question for future writing center tutors might be, How does this text use its processes and mechanics to build its argument? Writing centers are poised to continue serving as process-focused, procedurally literate hubs that support writing and composing practices across campuses, disciplines, and media—and this starts with seeing process as a key concept that binds together composition studies, writing center studies, and game studies.

While these may sound like lofty ambitions for writing centers, I return to the main thrust of this collection. The authors in this collection come to the topic of writing centers and game studies by asking "what if." But considering the many ways process brings writing studies, writing center studies, and game studies together, it encourages us to instead ask "how." Perhaps even importantly or playfully, it pushes us to ask "why not." One way to realize the potential for writing centers to continue serving as the "frontier of our discipline" is to focus our history of building on and engaging in processes. This almost certainly means asking more questions as well. Some of these questions may include asking how writing centers can support the digital and multimodal texts students increasingly produce as part of composition courses and across the curriculum, asking how the historically interdisciplinary writing center might continue to support increasingly digital work done across the university, and finally, given the incredibly interdisciplinary nature of game development, questioning how writing centers can build on our process-intensive history and knack for serving as a willing audience to better support this increasingly popular area of study.

Bogost's descriptions of what procedural literacy looks like in practice mirrors the daily practices of many writing centers surprisingly well, and thinking about how writing centers can encourage play in the context of both ideas and writing—to help students become better writers—demonstrates how writing center scholars can continue to build on the things at which writing centers excel. Throughout the past century, various uses of the term *process* have led to massive shifts in composition studies, writing center studies, and game studies; and a look across their various histories makes their past and present connections even more clear. More important, however, having demonstrated how process brings these areas together, I encourage writing center practitioners to consider how we can build on our rich history of process, encourage play in the writing center, and answer the difficult questions that arise when we start asking "what if" at the intersection of writing centers and games. The potential found in the answers seen throughout this collection—and those of anyone reading this text—can help keep writing centers at the frontier of our discipline.

REFERENCES

Almasy, Rudolph. 1982. "The Nature of Writing-Laboratory Instruction for the Developing Student." In *Tutoring Writing: A Sourcebook for Writing Labs*, edited by Muriel Harris, 13–20. Glenview, IL: Scott, Foresman.

Bain, Justin J. 2006. "Students and Authors in Writing Centers." In *Authorship in Composition Studies*, edited by Tracy Hamler Carrick and Rebecca Moore Howard, 75–88. Boston: Wadsworth.

Bannister-Wills, Linda. 1984. "Developing a Peer Tutoring Program." In *Writing Centers: Theory and Administration*, edited by Gary A. Olson, 182–96. Urbana, IL: National Council of Teachers of English.

Berlin, James A. 1987. *Rhetoric and Reality: Writing Instruction in American Colleges 1900–1985*. Carbondale: Southern Illinois University Press.

Bogost, Ian. 2006. *Unit Operations*. Cambridge, MA: MIT Press.

Bogost, Ian. 2007. *Persuasive Games: The Expressive Power of Videogames*. Cambridge, MA: MIT Press.

Bogost, Ian. 2008. "The Rhetoric of Video Games." In *The Ecology of Games: Connecting Youth, Games, and Learning*, edited by Katie Salen, 117–40. Cambridge, MA: MIT Press.

Boquet, Elizabeth H., and Neal Lerner. 2008. "After 'The Idea of a Writing Center.'" *College English* 71, no. 2: 170–89.

Braddock, Richard. 1974. "The Frequency and Placement of Topic Sentences in Expository Prose." *Research in the Teaching of English* 8, no. 3 (1974): 287–302.

Brannon, Lil, and Stephen North. 1980. "From the Editors." *Writing Center Journal* 1, no. 1: 1–3.

Brostoff, Anita. 1982. "The Writing Conference: Foundations." In *Tutoring Writing: A Sourcebook for Writing Labs*, edited by Muriel Harris, 21–26. Glenview, IL: Scott, Foresman.

Brown, Alan. 1990. "Coping with Computers in the Writing Center." *Writing Lab Newsletter* 15, no. 4: 13–15.

Carino, Peter. 1998. "Computers in the Writing Center: A Cautionary History." In *Wiring the Writing Center*, edited by Eric Hobson, 171–93. Logan: Utah State University Press.

Emig, Janet. 1971. *The Composing Processes of Twelfth Graders*. Urbana, IL: National Council of Teachers of English.

Faigley, Lester. 1986. "Competing Theories of Process: A Critique and a Proposal." *College English* 48, no. 6: 527–42.

Faigley, Lester, Robert D. Cherry, David A. Jolliffe, and Anna M. Skinner. 1993. *Assessing Writers' Knowledge and Processes of Composing*. Norwood, NJ: Ablex.

Flower, Linda, and John R. Hayes. 1977. "Problem-Solving Strategies and the Writing Process." *College English* 39, no. 4: 449–61.

Flower, Linda, and John R. Hayes. 1981. "A Cognitive Process Theory of Writing." *College Composition and Communication* 32, no. 4: 365–87.

Freedman, Aviva. 1982. "A Theoretic Context for the Writing Lab." In *Tutoring Writing: A Sourcebook for Writing Labs*, edited by Muriel Harris, 2–12. Glenview, IL: Scott, Foresman.

Garrett, Marvin P. 1982. "Toward a Delicate Balance." In *Tutoring Writing: A Sourcebook for Writing Labs*, edited by Muriel Harris, 94–100. Glenview, IL: Scott, Foresman.

Gere, John. 1985. "Empirical Research in Composition." In *Perspectives on Research and Scholarship in Composition*, edited by Ben W. McClelland and Timothy R. Donovan, 110–24. New York: Modern Language Association of America.

Harris, Muriel. 1982. *Tutoring Writing: A Sourcebook for Writing Labs*. Glenview, IL: Scott, Foresman.

Holmes, Leigh Howard. 1985. "Expanding the Turf: Rationales for Computers in Writing Labs." *Writing Lab Newsletter* 9, no. 10: 13–14.

Howard, Rebecca Moore. 2006. "The Binaries of Authorship." In *Authorship in Composition Studies*, edited by Tracy Hamler Carrick and Rebecca Moore Howard, 1–12. Boston: Wadsworth.

Kress, Gunther. 2010. *Multimodality: A Social Semiotic Approach to Contemporary Communication*. London, UK: Routledge.

Mateas, Michael. 2005. "Procedural Literacy: Educating the New Media Practitioner." *On the Horizon* 13, no. 2: 101–2.

Murphy, Christina, and Joe Law. 2000. "The Writing Center and the Politics of Separation: The Writing Process Movement's Dubious Legacy." *Contributions to the Study of Education* 79: 65–78.

Murray, Donald. 1976. "Teach Writing as a Process, Not Product." In *Rhetoric and Composition*, edited by Richard L. Graves, 79–82. Rochelle Park, NJ: Hayden.

Murray, Janet H. 1997. *Hamlet on the Holodeck: The Future of Narrative in Cyberspace*. Cambridge, MA: MIT Press.

Nash, Thomas. 1984. "Derrida's 'Play' and Prewriting for the Laboratory." In *Writing Centers: Theory and Administration*, edited by Gary A. Olson, 182–96. Urbana, IL: National Council of Teachers of English.

North, Stephen M. 1984. "The Idea of a Writing Center." *College English* 46, no. 5: 433–46.

Palmquist, Mike. 2003. "A Brief History of Computer Support for Writing Centers and Writing-across-the-Curriculum Programs." *Computers and Composition* 20, no. 4: 395–413.

Posey, Evelyn. 1986. "An Ongoing Tutor-Training Program." *Writing Center Journal* 6, no. 2: 29–35.

Rohman, D. Gordon, and Albert O. Wlecke. 1964. "Pre-Writing: The Construction and Applications of Models for Concept Formation in Writing." Cooperative Research Project 2174. Washington, DC: US Office of Education.

Shaughnessy, Mina. 1977. *Errors and Expectations: A Guide for the Teacher of Basic Writing*. Oxford: Oxford University Press.

Sheil, B. A. 1980. "Teaching Procedural Literacy." Paper presented at the Association for Computing Machinery Annual Conference, Palo Alto, CA, October 27–29. Abstract retrieved from https://dl.acm.org/doi/pdf/10.1145/800176.809944.

Sheridan, David M., and James A. Inman. 2010. *Multiliteracy Centers: Writing Center Work, New Media, and Multimodal Rhetoric*. Cresskill, NJ: Hampton.

Steward, Joyce S., and Mary K. Croft. 1982. *The Writing Laboratory: Organization, Management, and Methods*. Glenview, IL: Scott, Foresman.

Summerfield, Judith. 1988. "Writing Centers: A Long View." *Writing Center Journal* 8, no. 2: 3–9.

Tobin, Lad. 1994. "Introduction: How the Writing Process Was Born—and Other Conversion Narratives." In *Taking Stock: The Writing Process Movement in the '90s*, edited by Lad Tobin and Thomas Newkirk, 1–14. Portsmouth, NH: Heinemann.

4

READY WRITER TWO
Making Writing Multiplayer

Elizabeth Caravella and Veronica Garrison-Joyner

Questions about who has a right to language, translating between languages, language barriers, and respecting an individual's home language are significant considerations in writing center practice and theory (Boquet and Lerner 2008; Matsuda 2012; Salem 2014). As writing centers have begun to see a shift to multiliteracy centers in writing center work, writing center professionals continue to discover the relationships between a writer's social, cultural, historical, and technological experiences and their ability to make meaning in civic, academic, and professional contexts. Building on this premise and the work in composition pedagogy and gaming (Alexander 2009; Selfe 2010; Shultz Colby 2017), our chapter explores ways to adopt the language of video games in multiliteracy centers to explicitly interrogate and dramatically restructure the parameters of multicultural discourse in writing center practice by making otherwise opaque connections between policy and power transparent.

By examining the intersections of third space and possibility space from multiliteracy and gaming scholarship, respectively (Bogost 2011; Gee 2008; Gutiérrez, Rymes, and Larson 1995; Penuel and O'Connor 2018), this chapter illustrates how gaming concepts and language can be used in multiliteracy centers. More specifically, these concepts can help writing centers, especially multiliteracy centers, empower tutors and students. Putting these theories into practice gives writing center practitioners an opportunity to explicitly show writers how their interactions with the "rules" of writing assignments and classroom instruction function akin to a video game's feedback loop. Making these similarities visible helps students see writing as a recursive practice, much in the same way completing challenges and overcoming obstacles in a game are also recursive practices. We argue that this process makes writing

https://doi.org/10.7330/9781646421947.c004

"multiplayer" by restructuring discourse so that it allows for culturally inclusive language and writing styles: once writers become aware of the procedures and structures that govern their writing, they are in a better position to confront, navigate, and in some cases even change these rules.

First, we define a multiliteracy center and then establish how its design aligns with those used in gameful design. After establishing this connection, we delve deeper into the foundations of gameful design as illustrated through their connection to procedural rhetoric and possibility spaces. Finally, we connect possibility space with the concept of third space, synthesizing these theories to illustrate how bringing gaming language into multiliteracy centers helps writers both cultivate their own intrinsic motivations for writing and communicating ideas and internalize writing as a recursive and even experimental process.

THE MULTILITERACY CENTER

A term established by the New London Group (NLG) in 1996, multiliteracies refers to the multiplicity of ways meaning can be made. Recognizing that advances in technology were expanding the scope of available resources and shrinking the distance between the producers and consumers of those resources, the NLG initiated a path toward greater integration among disciplines. To that effect, the NLG asserts that to ensure equitable access to information and resources amid an increasingly globalized society, educators need to embrace the diversity of modal and linguistic tools available.

Moving from what it refers to as the "what" of multiliteracies to the "how," the NLG (1996) rests its argument on the concept of design, explaining that literacy pedagogues must understand themselves and their students "as both inheritors of patterns and conventions of meaning" and "active designers of meaning" (65). The multiliterate communicator layers linguistic, visual, audio, gestural, and spatial meanings together in multimodal patterns to produce a dynamic but unified message. The NLG challenges literacy educators in particular to prepare students to be designers of their own "social futures" through multiliteracy pedagogy that involves situated practice, overt instruction, critical framing, and transformed practice (66).

While the NLG (1996) makes no mention of writing centers in its work, Jackie Grutsch McKinney (2010) suggests that writing center scholars see their centers' potential reflected in the NLG's call for literacy education that empowers students to take control of their own writing. John Trimbur (2000) recounts the motivations behind changing his facility's name from the Writing Center to the Center for Communication across the Curriculum as a move to "signify the center's commitment not just to writing but to multiliteracies" (88) and sees his experience as indicative of a new turn in writing center theory and practice. Trimbur predicts, and we agree, that writing centers will continue to evolve into multiliteracy centers, in practice if not in name. He further notes that multiliteracy pedagogy in writing center contexts means a shift away from the perception of writing processes as internal, cerebral activities to be translated into words. Instead, a multiliteracies perspective sees the notion of design as "a production of visible language" (89). In other words, multiliteracy centers present an opportunity for students and tutors to lay the business of communication out in full view and critique it if moved to do so.

In the introduction to *Multiliteracy Centers: Writing Center Work, New Media, and Multimodal Rhetoric*, David M. Sheridan (2010) paints a picture of the "ideal multiliteracy center" as one that expands the writing center to include ample space and the tools to create multimodal compositions, staff who are equipped with rhetorical and technical knowhow, and a positive organizational disposition toward critical reflection (6). Sheridan and James A. Inman's (2010) collection of essays on multiliteracy centers strays further away from the NLG's original discussion of multiliteracy pedagogy than does Trimbur, moving instead toward conversations about multimodal composition and new media. Nevertheless, the understanding that we must make room for critical examination of the values that undergird notions of rhetorically effective and stylistically appropriate discourse remains.

In the same collection, Christina Murphy and Lory Hawkes (2010) assert that writing center professionals' developing role as interpreters of social architectonics is simply a continuation of writing centers' time-honored tradition of being agents of change. The architectonics, or arrangement of understandings that give structure to daily social activity, have as profound an influence on what a writer can reasonably do in a given text as the assignment parameters. In each multiliteracy center interaction, writers work collaboratively to understand not only the rhetorical situation in which they find themselves but the larger social framework that defines them in relation to their work and their institution. Murphy and Hawkes (2010) argue that multiliteracy centers are the natural evolution of writing centers in response to a smaller, more technologically mediated world. We would add that because they encourage multiplayer composition with specialists trained to explore the larger issues surrounding a text, multiliteracy centers also provide the space, time, and language for interrogating and perhaps even transforming inherited patterns of practice, especially when coupled with the principles of gameful design and procedural rhetoric.

GAMEFUL DESIGN AND MULTILITERACY CENTERS

As writing centers continue to evolve, their proponents must ensure that such an evolution remains inclusive. Adopting the language of video games within these spaces, especially in the case of multiliteracy centers, provides one such avenue. That being said, rather than rely on the usual practice of adopting gaming concepts in educational spaces known as gamification, we argue for an approach that relies on Jane McGonigal's (2016) conceptualization of gameful design.[1] As she describes in "I'm Not Playful, I'm Gameful" (2014), being gameful means "having the positive traits of a gamer," such as being goal-oriented and self-motivated, having self-confidence in one's abilities, actively pursuing new challenges, persevering through obstacles, and wanting to learn new skills (653). In other words, gamefulness relies on the intrinsic values playing video games can help cultivate rather than extrinsic ones, such as obtaining a particular item or defeating another team. Further, in looking at these traits, it becomes immediately recognizable that being gameful means being forward-thinking: there's always a new goal to reach, a new strategy to try, or a new skill to master. It makes sense, then, that the language used in and around games constructs them as always moving forward; by adopting such language in multiliteracy centers, we can better embody and instill that consistent forward momentum in the writers who visit us.

Unlike gamification, gameful design focuses on the individual and their ability to do what *they* want to do. An explicit example of this is the game app *SuperBetter*, designed and produced by McGonigal (2016). This game focuses on "building resilience" in its players, positing that "playing *SuperBetter* unlocks heroic potential to overcome tough situations and achieve goals that matter most" through gameful design rather than gamification (McGonigal 2016, 2). Concerned with the reduction of game design as merely a Skinner box exploiting operant conditioning, *SuperBetter* blogger Chelsea (2015) writes that "these gamification experts extolled all the superficial, short-term psychological hooks from games and none of the meaningful, metaphysical joy and satisfaction produced from playing. *They forgot that players are people*" (original emphasis). Continuing on to describe the key differences between gamification and gameful design, Chelsea highlights the key elements of gameful design that we cling to in our video game–based writing pedagogies: the desire to instill *intrinsic motivation* in writers rather than having them write for a grade or an extrinsic, less meaningful, and therefore less permanent motivation. In other words, gameful design works to change a player's *disposition*, or overall character, through practice. As McGonigal (2016) writes of *SuperBetter*, "Instead of taking the psychological hooks and operant conditioning from games, we use their deeply satisfying properties—things like agency, emotion, and immediate feedback—to help people do what they really want to do: feel better, reach their goals, connect with others, and live with meaning. We call this a *gameful* approach to design" (original emphasis).

This is not to say that all aspects of gamification are negative; in fact, utilizing grading scales that provide incremental rather than detrimental point earnings (in the same vein as "pointsification") or treating assignments and other course objectives as quests and achievements can be beneficial to the learning process and the overall design of the course, *if and only if these elements are incorporated with gamefulness in mind.* Through this lens, we can ensure that adopting the language of gaming not only benefits writers in their immediate classes and writing tasks but also has the potential to help them develop the intrinsic motivation necessary to continue writing well beyond the scope of a single semester. As well, this lens provides a means to help students develop their own sense of agency, perseverance, and curiosity.

Can a shift in language really have that big an impact? Thinking back to one of the most cited pieces about the power of language, George Orwell's (2013) *Politics and the English Language* argues that language operates as an instrument that can both shape the world for our own purposes and, in turn, shape our understanding of the world around us. In the case of bringing gaming language into multiliteracy centers, the words and phrases used in gaming can shape a place of openness, of possibility: there is always a way to progress in a game; you just have to figure it out. In addition, because video games are multimodal, both they and the language used to describe them overlap with the kinds of media projects multiliteracy centers help students create. By adopting language that forefronts capability, perseverance, and the idea that any player/writer can accomplish their goal, we open up these centers through language to bolster students' self-confidence and agency in their writing, especially when multiliteracy centers rely on gameful design for their policies and procedures. The writer who comes in with a paper that received a poor grade, for example, may feel downtrodden or otherwise defeated—feelings that can inhibit their ability to revise and resubmit.

Even something as simple as framing such papers within the language of games, where every "game over" comes with the option to "continue" or try again, can help combat writers' internalizing of a poor writing assignment (meaning they believe they are a poor writer). That is, especially because writing so often offers the opportunity to revise, we can use this language to frame revision as choosing that "continue" option.

Working from a gameful design standpoint in multiliteracy centers requires more than simply bringing gaming language into a writing center in an attempt to garner engagement. The way games use multimodal cueing to help players overcome obstacles can both serve as an example for students working on multimodal projects as well as provide language for tutors to use that may be more readily understood by their clients. As McGonigal (2012) suggests, then, video games that rely on gameful design create environments and experiences that cultivate these qualities in their players through the repetitive practice that occurs in play; for multiliteracy and writing center practitioners to do the same, this means incorporating not only the language of games and gaming but how this language reflects the underlying (and often invisible) design elements as well, which requires an understanding of how the procedurality of video games intertwines with the language that surrounds them.

PROCEDURAL RHETORIC AND GAMEFUL DESIGN

Games that employ careful attention to gameful design rely heavily on the ways the repetition and ritual of play through mechanics connect with the games as a whole. That is, gameful games do not separate their content from the programming that controls it; according to Ian Bogost (2008), that is a feat accomplished through procedural rhetoric. In *Persuasive Games*, Bogost (2007) defines the term *procedural rhetoric* as "the practice of using processes persuasively" (3). Although he used the term specifically to explain how video games construct arguments through procedures, procedural rhetoric, as well as the concept of "possibility space," or the "myriad of configurations" a player can access within a set of constraints, has great potential for developing a multiliteracy center from a gameful design perspective (121). For Bogost, the term *procedural* refers to the "established, entrenched ways of doing things" (3). In other words, procedures are both explicit and implicit rules or constraints that govern or influence our actions or the "processes [that] define the way things work: the methods, techniques, and logics that drive the operation of systems" (3). Although Bogost notes that procedurality can be examined in both "computational and non-computational structures . . . computational procedurality places a greater emphasis on the expressive capacity afforded by rules of execution" (4–5). What this means is that computer processes can show us how real-world processes can both restrict and encourage expression.

For example, as Rena Bivens (2017) discusses, on the surface, Facebook allows users to self-identify as a number of different genders, including "non-binary" and "genderfluid," purporting an illusion of diversity and inclusion. However, an examination of the proceduralized code behind these selections reveals that for the purposes of its marketing algorithms, Facebook only actually allows for three genders: male, female, or other (2017). By looking at the actual processes behind the user interface, Bivens (2017) illustrates how procedural rhetoric can illuminate otherwise hidden power structures because it brings attention to often invisible or otherwise black-boxed expressions of heteronormative power structures.

From this standpoint, procedural rhetoric allows us to see that despite the fact that we all act within certain restraints or processes, these processes often remain unnoticed or run in the background. The idea that we tend to only see a process "when we challenge it" (Bogost 2007, 45) illustrates these constraints as often opaque or black-boxed; that is, it is usually only when we seek to violate procedures that we notice them. This proves especially true for video games, as video game procedures are frequently black-boxed or run in the background through their programming, influencing players from a distance and only becoming visible when or if a player attempts to do something outside of those constraints. By reconceptualizing multiliteracy centers from a gameful design perspective, then, we can use procedural rhetoric as a means to assess and even revise some of the implicit power systems that would otherwise exude their influence behind the scenes by mapping out the possibility spaces within which different assignments, strategies, and even the centers themselves exist.

Tutoring Sessions as Possibility Spaces

After establishing his definition of procedural rhetoric, Bogost (2007) goes on to explain his conceptualization of possibility space. Relying on the concept of play as defined by Katie Salen and Eric Zimmerman, Bogost positions possibility space as the space "created by . . . processes themselves" (42). He explains that "in a procedural representation like a videogame, the possibility space refers to the myriad configurations the player might construct to see the ways the processes inscribed in the system work" (Bogost 2007, 42). However, Bogost's theory of procedural rhetoric only takes into account the perspective of video game designers. Miguel Sicart (2005), in contrast, argues that one must also examine the role players take when acting within the possibility space permitted by a game. As such, for the purposes of a gamefully designed multiliteracy center, rhetoric must be theorized from both the tutors' and the writers' perspectives, as the interactions between the two create the possibility space for each individual session.

Each individual session constitutes its own possibility space, though these spaces may be governed by larger center-wide or perhaps even campus-wide procedures. Because it is ultimately the tutors' and writers' interactions that define the space, each session—even those between the same tutor and writer—creates its own unique space. Within this space, then, lie the myriad configurations, or possibilities, available in a given session. With regard to writing, the concept of possibility space helps establish the idea of writing as a process, with no set rules or steps to follow but instead a space in which many moves are equally possible, though not all moves are equally effective. That is, much like a game, where all of the player's moves are readily available, the particular situation the player finds themselves in will dictate which moves might be more or less effective in that particular instance. In other words, any time a writer is responding to a writing assignment or even any time anyone writes in general, they are, in a way, operating within the possibility space as dictated by their stakeholders, audience, assignment prompt, and so on—the same way players experience the possibility spaces constructed by a particular game or individual level.

By adopting the concept of possibility space, tutors become better equipped to explain these ideas with language writers are more likely to be familiar with, since such concepts rely heavily on their connection with games and gaming. That is, not all writers may be practiced in thinking about writing in terms of a rhetorical situation or a writing situation, but most traditional college-age people are familiar with the term *level* as it's used in a video game. We can then use the term *level* (and the boundaries it creates with regard to play) as an analogy to jargon used with regard to the writing process, helping writers understand these concepts through ideas they're most likely more familiar with and possibly even more practiced with using and conceptualizing.

As discussed above, the language of video games is particularly forward-focused; even things that signify failure or the end, like a game-over screen, normally provide the option to continue or try again, as mentioned in the example above regarding revision. Furthermore, the idea of a "new game+" or playing the same game again after it has been beaten (often making the levels or bosses more challenging) mirrors the recursive nature of writing that novice writers frequently resist. Thus, although tutors are often already equipped to explain the content or rationale behind these concepts, purposefully using the same forward-functioning language helps influence writers to be forward-focused themselves. Sometimes, traditional language like revision or restructuring can seem daunting: it implies a heavy amount of intellectual work left to do. Similar to the above conceptualization of a rhetorical situation as a particular level, reframing these processes with the often more familiar and inviting language of games may help them seem like less of an obstacle and more of an opportunity.

Possibility Space and Procedurality in Practice

Although Bogost is working specifically within the realm of computational procedurality (that is, rules created by programs and algorithms), we argue that procedural rhetoric occurs in individual sessions as well through both human and nonhuman actors. Much like the black boxes created by these algorithms, teachers often set implicit rules and constraints for students to follow, and a portion of some sessions often involves explicating some of these boundaries. As noted by Jeff Sommers (2011) in his discussion of student reflections, reflection makes apparent "the influence on student reflection of the writers' desire to please their teachers" (100). In other words, tutors become an additional part of students' rhetorical situation, one that can operate as a type of tutorial, as within a game. Much like tutors, tutorials clearly explain and explicate the mechanics of a game so the players can then mix and match those mechanics for their own ends. Regardless of the mechanics, though, the players are still constrained by the rules of the game, though they are often encouraged to experiment or push against those rules. However, unlike in a video game, students rarely have the opportunity to challenge the procedures, both implicit and explicit, in a classroom, though they may try to resist these restraints through methods such as altering font sizes or making periods slightly larger to increase the number of pages in an essay.

Often, these procedural challenges manifest in the form of frustration with either an assignment or an instructor; they may appear to an instructor that the student is missing the point of a given assignment. The tutorial/tutor's role, then, is to connect the mechanics/process with the game/assignment; considering the role of the tutor in this way, especially for writers, helps position the tutor's role in the writing process in a way any game player would readily understand, as the games do not offer tutorials throughout their entirety since their purpose is to help you understand how to play rather than to play the game for you. In this way, establishing tutors as taking on that tutorial role helps prevent the tutor from becoming an additional point of resistance the writer must work through and instead frames them as a guide toward an effective solution rather than the solution itself.

In addition to more clearly establishing tutors as guides rather than solutions, framing tutors as points of tutorial method also carries with it implications that the term *tutor* does not; that is, often, many writers are reluctant to disagree with a tutor or to enact strategies not suggested by their tutor. These sites of resistance, though, from a gameful design perspective, can actually help pinpoint areas where either the student or the instructor has unmet expectations, and tutors can use this to help explain those implicit expectations. Adopting the language and procedurality of games within the center itself also gives the space to address some of this resistance more directly, helping create more productive sessions by allowing frustration to elicit a motivation in ways similar to the way video games are "pleasantly frustrating" (Gee 2008, 36). In video games, players know that some parts of a tutorial are not for them or are unnecessary because of their previous experience. As such, the tutor as tutorial method can help writers more quickly see that they are in control of their writing, even when they are working with a tutor they believe has more authority than they do. Tutorials do not dictate what a player can and cannot do; they merely make suggestions for what moves might be most effective in a particular possibility space. This language shift, then, can help mitigate resistance, as writers familiar with the concept of tutorials will be able to more easily embrace the tutor's role in their writing process without them feeling like the tutor has the ability and right to take over their piece.

By both understanding the writing center as a proceduralized space and conceptualizing individual sessions as their own unique possibility spaces, we set the foundation for adopting the language of video games from a gameful design perspective. By adopting gaming language in these spaces, writers become more aware of the implicit expectations established by an assignment sheet; in other words, the adoption of this kind of language begins to open the black box of the writing process. In summary, proceduralizing the multiliteracy center through the concept of possibility space can help us reconceptualize failure or resistance in students as well as better assess how implicit and explicit expectations in assignments shape and influence student response and engagement.

It also gives tutors the space to discuss assignments in terms of what can be done within an established realm of constraints while still positively framing these constraints. That is, the language of gaming can help tutors navigate student resistance to assignment prompts without having to comment on the prompts themselves. Outlining the possibility space afforded by an assignment, for instance, does not require making any judgments about the assignment, and relying on gaming language to discuss student resistance still leaves tutors in the realm of forward-focused feedback; even if students feel they have less expertise than their tutor, the positioning of tutor as tutorial helps establish the tutor's role as guide and helper rather than an authority they must obey in order to write well.

POSSIBILITY SPACE MEETS THIRD SPACE

In literacy studies, the concept of a third space presents a similar site of procedural rhetoric, wherein students and instructors, by purposefully interrogating their socio-historical positionality in relation to one another, expand what counts as knowledge. Kris Gutiérrez, Betsy Rymes, and Joanne Larson (1995) define the third space as a transitional place "where the possibility is created for both teachers and students to contest the transcendent script of the larger society" (452). They identify the scripts and counterscripts enacted by teachers and students, respectively, and describe the juncture where those scripts collide and create the potential for a renegotiation between parties. The third space is the potential that arises from that interaction, the potential to redefine what counts as knowledge.

In the third space, participants both test the limits of valid or appropriate discourse and potentially reshape those limits. Gutiérrez, Rymes, and Larson (1995) study the third space in a classroom context where the "teacher script" and "student counterscript" (445) exist in predefined, hierarchical relation to each other. The perceived immutability of those roles makes achieving the third space more of a challenge in a classroom, which Gutiérrez (2008) admits. Writing center spaces purposefully try to avoid imposing any kind of authoritative hierarchy, a value that inspires the policies of not discussing grades with students and not commenting on a teacher's assignment or practices. Tutors are peers who are literate in the official teacher script, but perhaps because they are not yet so entrenched in the role of instructor, the script seems less a part of their professional identities. Because students are encouraged to take agency over their work, writing centers are a prime location for habitual engagement in the third space.

Of course, writing center scholars are not ignorant of the third space. Bonnie Sunstein (1998) describes writing centers as "moveable feasts" (7), recognizing the ways writing center activity is constantly pooling in the liminal spaces between genres, the internal and external, the writer and reader, the writing center and larger campus community, school life and home life. What occurs in sessions must draw from so many disparate influences that it becomes necessarily adaptable, inescapably in-between. Third space represents a kind of in-betweenness in that it focuses attention on the boundaries between states of being. The writing studio is another realm where third space is prominent, although writing studio pedagogy (WSP) does not necessarily need to incorporate digital technologies.

In their foundational work on third space, Rhonda Grego and Nancy Thompson (2008) invoke the third space when describing the potentiality of writing studio work. They describe a place where students exercise "metarhetorical awareness of their higher educational institutional scene" and its effects on what counts as "knowledge" or "credible" (71). Grego and Thompson explain the dueling pentadic contexts vying for the author's consideration: the internal rhetorical situation of the text itself and the external, institutional forces that act on the enterprise of writing as student. The third space is created by the void between those sometimes competing realities. It is also the place where a student writer has the most agency because it is a space anchored in their individual experience. Students in a writing studio bring their writing projects and problems as well as personal concerns into a space where peers and specialists can provide insight and feedback. In this way, the writing studio supports multiplayer engagement in the third space by opening a student's individual writing process for collaborative intervention.

When we shift from writing center contexts to multiliteracy centers, we must dedicate even more energy to the larger, interdisciplinary, intertextual, and socially constructed dimensions of meaning making. Multiliteracy pedagogy, when applied by writing center professionals, should involve an ecological approach to composing processes. The move from a focus on individual writing processes to understanding the networked systems of interlocking processes that move like discordant clockwork requires deliberate interrogation of a given possibility space. Further, because video games are understood as an intersection of titles and their paratexts, or online writing about and around a game but not necessarily in the game itself (Paul 2010), adopting the language of gaming provides a readily understood foundation that positions writing and its processes as the same kind of interconnected network of scholars, texts, and ideas. Unearthing these connections cannot happen by accident but must be intentionally practiced at every opportunity. In short, while writing center interactions always have the potential to engage in the third space, in a multiliteracy center it is the main venue of discourse, and the concept of possibility space in particular provides a means for this kind of deliberate explication of constraints in a way other frameworks do not.

We imagine the third space as what Richard Selfe (2010) seeks when he asks what " 'state of mind' MLC [multiliteracy center] workers might adopt as they survey the human, technical, and social terrains around them" (116). While experienced writing center professionals must often grapple with the socio-cultural realities of writing, because technologies are ever-present and involved in the production of texts, Selfe emphasizes that in MLCs, special attention is also paid to the technical, nonhuman components of the system. In other words, MLCs are sites of continual negotiation and renegotiation of their own possibility spaces, not only interrogating what is possible from within but also exploring how external forces affect what has been possible in the past and what could be possible in the future. What Selfe sees as a "state of mind," Gutiérrez describes as a space, but both authors attempt to describe the seemingly intangible connections between individual writing processes and the social, historical, technical, and material realities within which they occur. When multiliteracy scholars discuss the potential for multiliteracy pedagogy to shape social futures, they are referring to the power of uncovering (New London Group 1996; Peneul and O'Connor 2018; Trimbur 2000). Multiliteracy center interactions should involve uncovering the often hidden rules of engagement that shape the ways students are able to engage in meaning making with their teachers and classmates.

Further, these interactions should provide an opportunity for students to see their writing activity as a multiplayer enterprise so they can understand their positionality within an ecology of writing. Marilyn Cooper (1986) states that writing and communicating through language is an "essentially social" activity, "dependent on social structures and processes not only in their interpretive but also in their constructive phases" (366). While an ecological approach to writing is not the exclusive purview of multiliteracy pedagogy, within this paradigm we come to understand the third space as an MLC's perpetual state of being. The open and pluralistic nature of multiliteracy work expands the potential configurations of positionality within various intersecting realms, not so much in contrast to writing centers but as a broader interpretation of the work. Multiliteracy center pedagogy encourages constant consideration of both immediate, internal rhetorical concerns and the larger external and material influences on meaning making in civic, academic, and professional contexts. In identifying the technological and human influences on any given composing situation, students and multiliteracy center specialists engage in third space negotiations.

The third space is uncomfortable because as Gutiérrez (2008) explains, it acts as a kind of "zone of proximal development" (148) wherein the student and the MLC professional work collectively to reach the next level.[2] For MLCs, the next level involves an expanded sense of place—not just where that place is affixed within its ecology but also the ways that positionality may be adjusted for new circumstances. Thus the third space is achieved not only by uncovering the hidden machinery behind our "everyday" and "institutional" interactions but also in the learning that takes place as students move from one context to the next and from one medium to the next (151). Because the language of video games shifts writing skills to a move-set for writers and requires that these moves exist within a space of creative constraints, such framing makes it more apparent to writers that there are always outside forces influencing their work.

Conceptions of possibility space in particular provide a means for explicating constraints that may otherwise remain unnoticed, especially those that remain more implicit in nature, such as an instructor's personal ethos not matching the ethos of a written assignment. While students may be able to easily navigate between situations when those situations seem similar—as in the first few levels of a game—the more complex and higher-stakes the assignment becomes, the less likely the student is to be able to transfer knowledge from one situation or level to the next. The growth that occurs between the space where the student can go on their own and the space occupying what they can achieve in a session often happens in the third space. Again, whereas the possibility space defines the parameters of what actions can be taken from within a given context or game by working together to try different configurations and push up against the boundaries, third space refers to a position of power where students do more than test the limits: they renegotiate them or at least imagine that they might do so. In this way, if possibility space helps position writers as the player or agent within a given rhetorical situation or level, third space lets them imagine their role as game designer or creator—a process that may help them consider why a given instructor assigns a particular prompt. In other words, combining the concept of possibility space with third space helps writers get at the underlying code that creates a particular writing situation. A student remediating a previously completed researched argument essay into a video project can work with a tutor to examine both the project's new rhetorical situation and the larger context that dictates appropriate rhetorical moves for this setting. Once the often elided institutional expectations are identified, they can be interrogated. The student can decide whether using certain population-specific terms, regardless of perceived academic merit, is warranted for the project. They can take a calculated risk, one they might never have considered otherwise.

By involving students in the proceduralization of multiliteracy centers, we may also empower them to engage in the critical framing necessary for self-identification within an ecology of discourse communities. Such socio-critical literacies do not come naturally and thus need to be developed through practice. Multiliteracy centers and multiliteracy pedagogies facilitate that development by providing space, support, and opportunity for a collective third space that not only makes the rules of engagement more obvious but also invites participants to question how those rules came to be and how they might be changed. Engaging students in the proceduralization of their own multiliteracy center visit can be as simple as asking the student to reflect on their personal learning goals and map what areas of the center would best support those goals. Considering their own goals for an assignment provides the student with the opportunity to consider how those goals differ from institutional expectations and, when coupled with principles of gameful design, helps further develop their own intrinsic motivations for achieving these goals. Further, the student's position as Player 1 is confirmed: while the activity will involve both the student and their peer specialist, the student is in control of the parameters; the specialist is just there to participate and respond. The goal of the multimodal project, when it begins, and when the activity ends are all up to Player 1.

Betsy Bowen and Carl Whithaus (2013) define multimodal composing as that which "involves the conscious manipulation of the interaction among various sensory experiences" (7). Based on this definition, then, video games are particularly well situated to benefit multimodal classrooms as, much like Gunther R. Kress (2010) argues with regard to multimodal compositions, games are also social-semiotic in nature, requiring attention to the whole system in which they exist to fully understand and appreciate the levels of meaning that exist. In other words, video games function as visible, explicit representations of the contextual, temporal nature of composition and as such can be particularly useful to help students transfer that understanding to their own writing. Similar to the way video games provide authentic experiences through virtual environments, because these games have their own inherent rule systems, they provide concrete examples of how different rhetorical situations call for varying strategies. Those familiar with games know that using a hack-and-slash approach in a turn-based strategy game will not be effective, for example. As such, because video games require attention to both large-scale and small-scale details, through gameful design, tutors can make the interactions of the global and local scales more explicit, as well as illustrate how such games reflect the writing process—all while using language with which younger writers especially are more likely to be familiar.

Responding to the need for writing to take on multiple modes and new genres in the digital age, multimodal composition practices and pedagogies continue to grow and expand as we find new uses and thus new needs for different types of media in our writing. As Kress (2010) points out, all types of multimodal communication require a social-semiotic approach to fully understand how a piece of communication exists for and within a specific context at a specific time and for a specific purpose. Similarly, Sheridan, Jim Ridolfo, and Anthony J. Michel (2012) amplify *kairos* as the key means of delivery for multimodal compositions, in that such compositions are more easily positioned within their existing contexts than are traditional written texts. By having a multiliteracy center founded in the principles of gameful design, the language of the center itself encompasses the kinds of ideas and projects writers may be working on when given multimodal assignments. This forward-focused language lays the groundwork for an examination of writing assignments and rhetorical situations in a way with which more writers will be familiar, one that positions them as the driving force and agent of change from the start. In this way, the writing center establishes its own ethos for these particular projects by embracing a common, shared experience that helps elicit a better understanding of multimodal communication practices and does so in a way specifically focused on unearthing non-inclusive practices that may be woven into the larger institution in which the writing center operates.

Especially when considering the forward focus of gaming language in particular, some may be concerned with bringing the kind of win-lose language gaming utilizes into writing center spaces. However, as previously mentioned, the concept of losing in a video game is much different than it is even in sports or other competitions. That is, video games position losing as learning; if the player fails to obtain an objective, what they have really done is explored a potential avenue that ended up being ineffective for that particular situation. In this way, by adopting the language of games, multiliteracy centers may be able to better address writing as a process rather than merely the product that receives an assessment. That is, gaming language allows us to understand that losing, rather than being an end, just illustrates what's less rhetorically effective. As such, when we adopt this kind of language, we give writers room to experiment and respond to feedback; the only time a game-over screen means a player cannot try again is if they themselves choose to quit.

MULTIPLAYER WRITING AND GAMEFUL DESIGN IN MULTILITERACY CENTERS

So far, most of what we've discussed in and around games and gaming in the multiliteracy center seems to focus only on the individual. However, by understanding these sessions as their own individual possibility spaces where tutors and writers come together to tackle a particular writing challenge, we can begin to illustrate how a gamefully designed center provides a framework for understanding writing as multiplayer. To begin, consider what can be done in a video game when switching from single player to multiplayer. *World of Warcraft* provides an excellent example, as players often do their initial leveling by themselves, or solo, but are only able to complete the end-game content by playing with others. Players who do not want to participate in multiplayer activities end up unable to obtain some of the game's best items and equipment. Those who want such rewards must understand how to work as part of a team because if one person fails to do their part in end-game content, it usually results in the death of the entire team.

When gaming shifts to multiplayer, it grants access to new levels or other activities that are inaccessible in single-player mode and often offers better rewards for overcoming these obstacles. In addition, the end-game content requires players to collaborate to develop effective strategies for a given obstacle within the game's provided boundaries (as all players must still operate within the game's established possibility space). In the same way, when writers work with a writing consultant or "digital content specialist" (Sheridan 2010, 3), both the writer and the tutor are making meaning and testing these boundaries together. Adopting both the language and design elements of a video game within the writing center, then, predisposes writers to the idea that collaboration and cooperation lead to greater rewards or in this case, greater writing ability.

By establishing the center as a multiliteracy center, the possibility spaces inhabited by both tutors and writers demand that we not only collaborate on what is possible but examine the socio-historical context that shapes the possibility space and imagine ways of expanding it. That is, because possibility space requires the explicit naming of limitations and constraints that create a given space, tutors and tutees can map these out together in a way that purposefully opens the black box. In this third space, where the student and multiliteracy center professional take an openly socio-critical stance in relation to both each other and their shared context, the collective enterprise of meaning making is laid bare. All participants can be deliberate about understanding and being understood, recognizing that there are more players involved than just one writer trying to win over an audience. When students shift from a single-player to a multiplayer mind-set, the result is a necessarily more dynamic sense of audience awareness. Consistent, deliberate engagement in the third space is an ideal that may require some share of the training space for specialists. Nevertheless, tutors and students frequently stumble into the third space whenever conversations swerve away from the project and into the challenges of being a college student or how to communicate comprehension issues to the professor. Think of how much students and specialists can do together when those moments are sought out.

By taking on the language of gaming from a gameful design perspective, we give writers the tools to not only improve their writing but also to question and uncover some of the hidden power structures operating behind the scenes in their institutions and professional lives. Proceduralizing the writing center makes these black-boxed power structures more visible, and adopting the language of video games helps empower these writers to confront and begin revising or removing these entrenched values so they can be inclusive not only in appearance (as with the Facebook example) but in practice as well, starting with the processes and procedures that govern them.

NOTES

1. For some of the common critiques/problematizations of gamification, see Bogost (2008).
2. For more on the zone of proximal development, see Vygotsky (1978, 157).

REFERENCES

Alexander, James. 2009. "Gaming, Student Literacies, and the Composition Classroom: Some Possibilities for Transformation." *College Composition and Communication* 61, no. 1: 35–63.

Bivens, Rena. 2017. "The Gender Binary Will Not Be Reprogrammed: Ten Years of Coding Gender on Facebook." *New Media and Society* 19, no. 1: 880–98.

Bogost, Ian. 2007. *Persuasive Games: The Expressive Power of Video Games*. Cambridge, MA: MIT Press.

Bogost, Ian. 2008. "The Rhetoric of Video Games." In *The Ecology of Games: Connecting Youth, Games, and Learning*, edited by Katie Salen, 117–40. Cambridge, MA: MIT Press.

Bogost, Ian. 2011. "Exploitationware." In *Rhetoric/Composition/Play through Video Games*, edited by Richard Colby, Matthew Johnson, and Rebekah Shultz Colby, 139–48. New York: Palgrave Macmillan.

Boquet, Elizabeth H., and Neal Lerner. 2008. "Reconsiderations: After 'The Idea of a Writing Center.'" *College English* 71, no. 2: 170–89.

Bowen, Betsy, and Carl Whithaus. 2013. *Multimodal Literacies and Emerging Genres*. Pittsburgh: University of Pittsburgh Press.

Chelsea. 2015. "Gameful Design." *SuperBetter Blog*, June 22. blog.superbetter.com/gameful-design/.

Cooper, Marilyn. 1986. "The Ecology of Writing." *College English* 48, no. 4: 364–75.

Gee, James Paul. 2008. *Good Video Games + Good Learning: Collected Essays on Video Games, Learning, and Literacy*. New York: Peter Lang.

Grego, Rhonda, and Nancy Thompson. 2008. *Teaching/Writing in Third Spaces: The Studio Approach*. Carbondale: Southern Illinois University Press.

Gutiérrez, Kris. 2008. "Developing a Sociocritical Literacy in the Third Space." *Reading Research Quarterly* 43, no. 2: 148–64.

Gutiérrez, Kris, Betsy Rymes, and Joanne Larson. 1995. "Script, Counterscript, and Underlife in the Classroom: James Brown versus *Brown v. Board of Education*." *Harvard Educational Review* 65, no. 3: 445–72.

Kress, Gunther R. 2010. *Multimodality: A Social Semiotic Approach to Contemporary Communication*. London: Routledge.

Matsuda, Aya. 2012. "Teaching Materials in EIL." In *Principles and Practices for Teaching English as an International Language*, edited by Lubna Alsagoff, Sandra McKay, Guangwei Hu, and Willy Renandya, 168–185. New York: Routledge.

McGonigal, Jane. 2012. *Reality Is Broken: Why Games Make Us Better and How They Can Change the World*. New York: Penguin.

McGonigal, Jane. 2014. "I'm Not Playful, I'm Gameful." In *The Gameful World*, edited by Steffen P. Walz and Sebastian Deterding, 653–58. Cambridge, MA: MIT Press.

McGonigal, Jane. 2016. *SuperBetter: The Power of Living Gamefully*. New York: Penguin.

McKinney, Jackie Grutsch. 2010. "The New Media (R)evolution: Multiple Models for Multilieracies." In *Multiliteracy Centers: Writing Center Work, New Media, and Multimodal Rhetoric*, edited by David M. Sheridan and James A. Inman, 207–24. Cresskill, NJ: Hampton.

Murphy, Christina, and Lory Hawkes. 2010. "The Future of Multiliteracy Centers in the E-World: An Exploration of Cultural Narratives and Cultural Transformations." In *Multiliteracy Centers: Writing Center Work, New Media, and Multimodal Rhetoric*, edited by David M. Sheridan and James A. Inman, 175–88. Cresskill, NJ: Hampton.

New London Group. 1996. "A Pedagogy of Multiliteracies: Designing Social Futures." *Harvard Educational Review* 66, no. 1: 60–93.

Orwell, George. 2013. *Politics and the English Language*. London: Penguin Classics.

Paul, Christopher. 2010. "Process, Paratexts, and Texts: Rhetorical Analysis and Virtual Worlds." *Journal of Virtual Worlds Research* 3, no. 1: 4–17.

Penuel, William R., and Kevin O'Connor. 2018. "From Designing to Organizing New Social Futures: Multiliteracies Pedagogies for Today." *Theory into Practice* 57, no. 1: 64–71.

Salem, Lori. 2014. "Opportunity and Transformation: How Writing Centers Are Positioned in the Political Landscape of Higher Education in the United States." *Writing Center Journal* 34, no. 1: 15–43.

Selfe, Richard. 2010. "Anticipating the Momentum of Cyborg Communicative Events." In *Multiliteracy Centers: Writing Center Work, New Media, and Multimodal Rhetoric,* edited by David M. Sheridan and James A. Inman, 109–29. Cresskill, NJ: Hampton.

Sheridan, David M. 2010. "Introduction." In *Multiliteracy Centers: Writing Center Work, New Media, and Multimodal Rhetoric,* edited by David M. Sheridan and James A. Inman, 1–17. Cresskill, NJ: Hampton.

Sheridan, David M., and James A. Inman, eds. 2010. *Multiliteracy Centers: Writing Center Work, New Media, and Multimodal Rhetoric.* Cresskill, NJ: Hampton.

Sheridan, David M., Jim Ridolfo, and Anthony J. Michel. 2012. *The Available Means of Persuasion: Mapping a Theory and Pedagogy of Multimodal Public Rhetoric.* Anderson, SC: Parlor.

Shultz Colby, Rebekah. 2017. "Game-Based Pedagogy in the Writing Classroom." *Computers and Composition* 43: 55–72.

Sicart, Miguel. 2005. "Game, Player, Ethics: A Virtue Ethics Approach to Computer Games." *International Review of Information Ethics* 4: 14–18.

Sommers, Jeff. 2011. "Reflection Revisited: The Class Collage." *Journal of Basic Writing* 30, no. 1: 99–129.

Sunstein, Bonnie. 1998. "Moveable Feasts, Liminal Spaces: Writing Centers and the State of In-Betweenness." *Writing Center Journal* 18, no. 2: 7–26.

Trimbur, John. 2000. "Multiliteracies, Social Futures, and Writing Centers." *Writing Center Journal* 20, no. 1: 88–91.

Vygotsky, Lev. 1978. *Mind in Society: The Development of Higher Psychological Processes.* Cambridge, MA: Harvard University Press.

PART 2

Applications of Games to the Writing Center

5

LEVELING UP WITH EMERGENT TUTORING

Exploring the Ludus and Paidia of Writing, Tutoring, and Augmented Reality

Brenta Blevins and Lindsay A. Sabatino

Several years ago, our writing center started encountering class assignments that required augmented reality (AR). These AR assignments ranged from static advertisements to interpretive labeling to informative videos. While our campus had resources for responding to student questions about using the multiple digital technologies required to create AR, we realized that our center had to develop tutor training and strategies for responding to the rhetorical effectiveness of AR texts. How could we as a writing center, well versed in tutoring traditional academic writing forms such as the essay and the report, tutor students who had composed texts in this new medium?

Around the same time AR started appearing in campus assignments, the mobile AR game *Pokémon Go* (https://nianticlabs.com/en/support/pokemongo/) was released in 2016. As discussed in this book's introduction, *Pokémon Go* is a game in which players explore their environments to collect fantasy monsters, navigate rules, and engage in multiple forms of play to acquire new knowledge and skills that they apply in continued game play. Unlike games that end when players complete a puzzle or confront a final challenge or opponent, *Pokémon Go* has no set end. *Pokémon Go* is a game that supports ongoing play without end and thus falls under the game-play classification of "emergence" (Juul 2005, 5). By contrast, games that end with a final task or "boss" challenge are called games of "progression" (5). Players are less likely to replay games of progression because they have seen the end of the game's story. By contrast, *Pokémon Go* is highly re-playable, as is characteristic of emergent games.

https://doi.org/10.7330/9781646421947.c005

While playing *Pokémon Go,* we also came to understand theory about different kinds of game play: *ludus,* which is structured, rule-bound play pertaining to the ultimate goal of winning a game, and *paidia,* which encompasses exploration and spontaneous play (Ang 2006; Caillois 1961; Frasca 1999; Tosca 2003). This game-play theory soon inspired the way we approached tutoring AR. To become a center equipped with tutors who could confidently address AR, we realized we would have to work within the rules of the new medium, but we'd also have to acquire new knowledge, skills, and approaches that could not be found through memorizing a list of rules about using AR. We would have to explore and continue to level up. In doing so, we developed and employed what we term an emergent theory of tutoring that relies on play, experimentation, and game-play approaches that combine structure and exploration.

We weren't the first to grapple with the problem of changing forms of writing. In recent years, writing centers have increasingly worked with students composing in a range of media (Carpenter and Lee 2016; Grouling and McKinney 2016; Trimbur 2000) and changing forms of composition (Yancey 2004). Such changes have raised questions for writing centers and tutors (Balester et al. 2012; McKinney 2009) about responding to writing outside more familiar academic genres and media. As more writers are producing multimodal texts, writing centers are finding ways to prepare tutors to respond through a range of pedagogies (Carpenter and Lee 2016). The negotiation and evolution of writing centers to respond to less traditional texts have been further outlined in Elizabeth Caravella and Veronica Garrison-Joyner's chapter in this volume, as well as in this book's introduction. Here, our chapter describes how we addressed the challenges of responding to new forms of writing, such as AR, by including game studies in the writing center. Our work builds on game-play theory previously proposed for writing centers (Lochman 1986) and also for writing classes (Alberti 2008; Colby 2017; Colby and Colby 2008; Gee 2003; Rouzie 2000; Sabatino 2014). Our theory of emergent tutoring draws inspiration from Rebekah Shultz Colby and Richard Colby's (2008) development of emergent pedagogy, which combines work with game play and promotes invention in the writing class. Our emergent tutoring further employs *ludus* and *paidia* as strategies for developing plans for and responding to new forms of multimodal writing.

PREPARING FOR AUGMENTED REALITY TUTORING

To us, tutoring AR and composing AR sounded daunting at first—a little scary, a bit like confronting a monster in a game. We needed to defeat our monster: the goal of tutoring AR while being uncertain how to proceed, which necessitated getting familiar with our new world. To get started with AR tutoring through emergent tutoring, we developed tutor training that was part *work*shop and part *play*shop, beginning with exploring—a key aspect of emergent tutoring. Game players need time to acclimate to the new world they're playing in, so we shared materials that described how AR is a medium in which digital content is combined with a user's surrounding environment, material objects, or both. The training described some of the rules and expectations around AR use. For example, an AR user might point their smartphone camera at an object to trigger AR text, video, or audio about that object. Such digital text might be shown in combination with a digital representation of the user's surroundings. For example, when a smartphone user activates AR software to look at a historic building, the AR digital layer might show on the screen explanatory text pointing to historical changes in the building. While AR has been around for decades (Arth et al. 2015), it has recently risen in awareness and availability through a combination of new and more affordable hardware and AR software.

Our training provided exploration of multiple AR examples. While AR became popular through gaming, it has numerous emerging uses, ranging from advertisements to museum interpretation to graffiti, among other projects. We talked about other sets of rules around AR composing, specifically around the assignments we saw increasingly in higher education. While some of these assignments appear in writing classes (Blevins 2018; Morris 2019; Tham et al. 2016), AR composing projects also appear across the curriculum in education, human resource development, marketing (Delello, McWhorter, and Camp 2015), and history (Förster and Metzger 2015; Greene 2018), among others. Because of this increase in AR composing assignments and growing use of AR in professional marketing, technical communication, and software apps, writing centers should prepare for AR tutoring.

Addressing a few concerns about whether the center should tutor AR, we cited how Michael Pemberton (2003) had previously discussed a different form of digital writing by observing that "even if most instructors will not encourage hypertext papers or teach Web design in their courses, others certainly will"; in fact, some may view "Web design and hypertext as integral parts of their curricula" (19). Further, while AR may not be a medium assigned in every writing class, it can appear in classes across the curriculum. Familiarity with AR tutoring provides benefits not just for that medium but also for engaging with proactive, forward-planning conversations around tutoring other multimodal texts that raise similar questions in a dynamic writing environment.

Our workshop wasn't designed to be all work. We also made time for play, such as *Pokémon Go.* A few of our tutors had already played the game—some enthusiastically—but several had not. The more experienced players helped the less experienced with the basics of the game and collaboratively explored the *Pokémon Go* world. The experienced players talked about how they were continuing to learn new aspects of *Pokémon Go* game play. After catching a few Pokémon, examining how the campus was represented in the game, and taking a few selfies with the AR Pokémon, we talked about different game-play concepts, such as emergence and progression. For example, *Pokémon Go* is an emergent game because it supported a wide range of ongoing game play and the ongoing learning of our experienced *Pokémon-Go*–playing tutors. That sparked a conversation about how both writing and tutoring are ongoing processes in which one may become a better writer or a better tutor but that both writers and tutors alike have more to learn. In that way, writing and tutoring are like games of emergence, highly re-playable, supporting ongoing exploration, and focused less on winning and more on playing.

Informed by Chee Siang Ang (2006), Roger Caillois (1961), Gonzalo Frasca (1999), and Susana Tosca (2003), we also talked about the two different forms of game play: *ludus* and *paidia.* Ludic game play focuses on structured, rule-oriented interactions that present the outcome of winning or losing the game. Ludic games have set rules for what constitutes winning, which can be accomplished through both intrinsic and extrinsic game play. As Ang (2006) explains, intrinsic ludic rules contribute indirectly to winning a game, whereas extrinsic ludic rules contribute directly. Usually, extrinsic ludic rules are stated explicitly in the game as the ultimate goal, whereas intrinsic rules need to be constructed by players or are introduced in the game from time to time to achieve the winning condition (311). While the player may not always choose to engage with ludic game play, the player may not be able to win a game without doing so (Ang 2006). These elements of ludic game play can be found in *Pokémon Go* in the game's tagline, "Catch 'em all!" Unlike games that focus on defeating a boss or retrieving a singular winning object, *Pokémon Go* is a collection-based game in which players level up as they continually collect the game's pocket monsters, even after they hypothetically "catch 'em all."

Play, in the sense of *paidia* (also spelled as *paidea*), emphasizes spontaneous, improvisational exchanges (Caillois 1961). *Paidia* game play does not have the structure of a winning scenario or set goal as defined in ludic games—for example, role-playing games, adventure games, and augmented reality interactions. Instead, "The player has the freedom to determine her goals . . . as soon as the *paidea* player determines a goal with winning and losing rules, the activity may become a *ludus*" (Frasca 1999). Therefore, *paidia* game play has a spontaneous player-driven structure and involves exploring the game. *Pokémon Go* supports *paidia* in several ways. First, because the game displays a map based on players' physical locations, players are free to choose when and where to go in the real world, and thus they can choose to explore both their real-world contexts and their *Pokémon Go* environment. In other words, *Pokémon Go* is an open-world game that allows players to determine where they want to go to collect the game's Pokémon. Every time they access the game, players are likewise free to decide if they want to focus on Pokémon collection, battle other Pokémon in gyms, participate in raids against boss-level Pokémon, collect certain types of Pokémon, or seek new Poké Stops to collect balls and other game-play items. Players constantly move back and forth between these two types of game play (*ludus* and *paidia*) and rule sets.

While *paidia* does not have the overall goal of winning, *paidia* offers a structure that involves the rules that operate the game by defining "what the players can and cannot do" (Ang 2006, 310). Differentiating the two types of play, Ang (2006) explains, "*Ludus* rules state what should be done, whereas *paidea* rules state how it should be done" (313). These rules are established by the game designer(s) determining how players can interact with the game and what they can or cannot do within the gaming environment. For example, in *Pokémon Go*, player avatars move across the game-play map to find Pokémon by walking but cannot teach their avatars to fly. When attempting to acquire Pokémon by collecting them in Pokéballs, players are limited to throwing either direct balls or curve balls at their targets; no other ball-throwing options are available to them.

After we finished our game-play theory discussion, we transitioned to discussions of tutoring AR and applying our understanding of game concepts. For example, a tutoring session engages the ludic rules of assignments, genres, media, purposes, and objectives, while *paidia* provokes exploration into the realms of possibility for a writing project—a navigation between the space of the writing center and the discourse community for which the project is intended. We then tried out AR composing software and planned for the sorts of challenges and questions we might experience in tutoring sessions, whether those might be for marketing, history, or composition assignments.

In the following section, we articulate our emergent tutoring process using game-play concepts and strategies we identified while applying emergent tutoring for AR. Although we focus primarily on AR composing, the game-play concepts of *ludus* and *paidia* and emergence and progression can be helpful in any consultation, whether for multimodal or traditional academic writing assignments.

RESPONDING TO AUGMENTED REALITY USING EMERGENT TUTORING

The structure of a writing center consultation session involves conversational interactions between writer and tutor that are similar to game-play rule negotiation. In a consultation, tutors and writers engage with both *ludus* and *paidia* game play. During the initial setting of a session agenda (Macauley 2005), a student might set the goal of getting an A on a writing project (a ludic-style "winning"); however, the tutor's *paidia*-focused role hones in on producing "better writers, not better writing" (North 1984, 438); thus the tutor can help the student reframe their initial *ludus*-oriented, grade-focused goal. As the tutor may not be an expert in the content or medium the student is working on, the tutor responds by employing the improvisation of *paidia*: asking questions, discussing options, and offering opportunities for trial and error—all strategies used in traditional consultations. The tutor and writer have different roles in sessions, which means there are times when the student is operating within ludic game play while the tutor is within *paidia* game play, and they employ flexibility in moving between the different forms of game play.

Building from a similar understanding of emergent and progression gaming (Juul 2005), Colby and Colby (2008) theorize an emergent pedagogy for the writing classroom in which "teachers introduce writing principles and strategies in order to open up a studio-like space for students to work through those strategies on their own" (305). Such a strategy resembles that of tutors—and *Pokémon Go*'s guide Professor Willow—who provide input that guides future composing by making space to focus on exploring options. Professor Willow is an onscreen character who occasionally appears in *Pokémon Go* to offer *paidia*-esque guidance but not explicitly ludic direction. For example, Professor Willow appears to issue players special challenges occasionally, greeting them with an encouraging message such as "Trainer, how are you? It's been a while, hasn't it? You've become quite the Trainer" (Niantic 2020). These special challenges are optional, suggested activities for the player: "This'll be a great time to take your buddy out to explore a bit and . . . complete the following research tasks" (Niantic 2020). However, it is up to the player whether to take up the tasks and in what manner; the player is left to figure out specific strategies on their own to carry out these tasks. Throughout, the *Pokémon Go* player generally drives the game, much as the writer controls the tutoring session, while Professor Willow provides a representation of peer tutors who from time to time encourage writers to play and experiment with their writing and writing processes rather than deliver explicit direction. In recognizing the similarities between responding to media and game play, tutors can apply their prior tutoring experience and with confidence guide writers composing in unfamiliar media into the session.

Beginning the Session

Every tutoring encounter starts when the writer initiates game play with the tutor by making an appointment or dropping into the center for a session. At that moment, the writer may already have a ludic goal in mind, a clear expectation of what they want to achieve. They also may be in the exploratory *paidia* framework, as they are looking for assistance but have not established a winning goal. Although the writer's ludic- and progression-focused goal of making an AR composition is important, the tutor can employ *paidia* to more deeply engage emergence in the AR composing process.

Building Rapport and Understanding the Assignment

Some students bring in new projects feeling intimidated, believing, for example, that AR is such a complicated or unfamiliar medium that it's going to defeat them. A good strategy for responding to these emotional states is to build rapport between tutor and writer. Establishing rapport helps students in gaining confidence, establishing trust, and creating a bond (Sabatino 2009). To do so, the tutor aims to understand the writer and learn more about them; they may ask questions about the writer—major, classes, interests, and the like. Building rapport is primarily *paidia*, as the writer and tutor get to know each other and establish an environment for learning. The tutor and writer explore together the writer's interests, which might inform the writer's focus or topic for the composition.

Rapport building might reveal that the student is having trouble with their AR composition because they've focused too much on *ludus*, trying to win by learning just enough AR to simply finish a project. The tutor can suggest a more emergent response: getting familiar with AR itself. The tutor and student might work to identify the range of capabilities of the AR composing software. Or the student might reveal that they are having trouble developing any material for the assignment because they are uncomfortable with AR composing. The tutor might ask the student what they know about AR. The two can spend time exploring the world of AR composing. If the student has technology with them—such as a phone or tablet and AR software—the tutor and student can examine invention options together. Discussing what is possible and finding models can help the student start imagining possibilities for their composition. The tutor might ask if the writer has experience with *Pokémon Go* or another AR; they might briefly look at an AR game or application during the session to discuss what they notice.

Ludus can also be a productive strategy for helping students who are frustrated by their AR assignments. For example, as rapport is built, the tutor and writer will move on to gain an understanding of the assignment, when it is due, and why the writer has activated the writing center as a resource within the game of becoming better writers. This stage continues in the realm of *paidia* game play as it moves closer to establishing the ludic/winning goal. The writer explains the purpose of the assignment and shares their understanding of what is being asked of them. If the writer has brought the assignment instructions, the tutor may also read through them. In some ways, this is a game within the tutoring session, or a mini-game, as the tutor and writer decode the rules together.

This negotiation of *ludus* and *paidia* during the beginning of the session means that the tutor and writer work together to develop an approach to the assignment. For example, the student might bring in a marketing AR assignment that requires the development of a print target image and a digital component to advertise a campus event of the student's interest. The tutor's questions might lead the student minoring in music to develop interest in creating an AR marketing campaign for an on-campus musical performance later in the semester.

Setting the Agenda and Determining the Goals of the Session

Although the first two session stages may use *ludus*, they are primarily *paidia*-style game play. Once the tutor asks about the writer's goals and why they sought assistance from the writing center, the session shifts toward *ludus*-related goal setting. The tutor may seek to clarify terms the writer is using, such as layout or flow, as a way to make sure they share the same understanding. This stage is where *ludus* gets established and refined, where the tutor enters the writer's *ludus* game. The writer might set the goal to gain a better understanding of how to effectively design an AR print advertisement and leave the session with a clear plan, with each step identified for moving forward with their AR marketing assignment.

The tutor then uses information gathered from those questions to collaboratively set an agenda, determining the overall goals for the session and drawing a map. Together, the tutor and writer explore the options of the writing project and tutoring session. The agenda that is created between the tutor and writer sets the shared ludic/winning goals. Macauley (2005) explains agenda setting this way: "Charting the course for a tutorial lesson is also a way to mark, simply and graphically, the things you want to do in the tutoring session: 'Begin the session with _____ then _____ and conclude by _____.' Fill in the blanks as you wish" (1). The agenda identifies options and allows both the writer and tutor to have shared, focused goals.

Setting the agenda is also a stage that continues to build the *paidia*-style rules. The tutor may explain what will happen in the session; for example, the tutor might identify that their job is not to create the AR composition for the writer but to provide feedback on the writer's ideas or designs. Together, they may also discuss who will take notes or type on the computer. The tutor will use this discussion and the information from building rapport to explain what is and is not possible in a session; for example, the tutor might point out that although the writer may want to leave with a functioning, fully designed AR, they hadn't begun the session with a topic identified and only scheduled a thirty-minute session. The tutor would use the *paidia* rules of time constraints to determine a realistic agenda.

While some rules such as time constraints are fixed, other rules, like taking notes or setting the agenda, are more malleable. As Lauren Fitzgerald and Melissa Ianetta (2016) explain, the tutoring session is flexible. Similarly, William Macauley (2005) compares setting the agenda to drawing a map and identifies that the advantage of such a map "is that it can be negotiated. When the map is negotiated, it's multivalent, meaning that both the tutor and writer can plan the tutorial cooperatively without either dominating the session or being tied unproductively to the writer's text" (3). Together, the tutor and writer negotiate priorities and determine what areas are most important to focus on, much like *Pokémon Go* players explore their in-game maps and focus on particular areas over others. That sort of collaboration in tutoring mirrors the collaborative game play we observed between experienced and novice *Pokémon Go* players. Jesper Juul (2002) notes that some aspects of cooperation can be found through emergence, for example, when teams of players need to work together to accomplish a task. For the AR marketing tutoring session, the tutor and the writer can work together to determine effective design elements for a printed trigger that will activate the display of the AR text and then devise a plan for developing on their own the digital side of the AR project when the student leaves the session.

Understanding the agenda and goal-setting portions of the session as rule negotiation provides an opportunity for recognizing how the familiar aspects of tutoring alphabetic academic projects apply similarly to tutoring multimodal projects. If a student brings in an AR or other digital composition, tutors might initially worry that they cannot return to their familiar strategy of reading aloud the writing, which is a more complicated activity when working with multimodal designs and digital media, such as websites or augmented reality. How does a tutor read aloud a visual? Or hyperlinks? However, tutors can employ the familiar rules of tutoring sessions to support a process-based approach that applies not just to traditional academic writing but also to responding to and composing writing in various media and using multiple modes.

By viewing sessions as shaped by tutoring conversational rules that ultimately foster the writer's development, the tutor can, to use Jackie Grutsch McKinney's (2009) term, "talk" through the text, "showing the student how it could be read in its entirety" (39). In a consultation for an essay, a tutor will often say something like "I will read the paper aloud and point out anything that stands out to me. Please do the same." An AR composition can involve the tutor taking on the role of the audience and talking through their experience of the AR; they might narrate their responses as they trigger the project.[1] Just as a tutor would explain before implementing the agenda of a traditional writing session that the tutor and writer might perhaps take turns reading the paper aloud, the tutor can help the writer in a multimodal session, such as an AR consultation, understand how they will talk through the composition. For the AR marketing assignment, the tutor and writer might brainstorm and sketch out a combination of text and images to draw attention to the trigger advertisement; instead of deciding on or drawing actual images during an initial phase, the tutor and writer might discuss the options for visual triggers and use placeholders for images the writer will decide on later.

Talking through a further developed AR project might mean that the tutor would start by asking the writer about the trigger that will initiate the AR composition for the audience. How does that trigger fit the content of the composition? How will the audience recognize the trigger within their everyday environment so they can activate their AR software? When a writer brings in a fully functioning AR project, the tutor might then discuss what they notice in the video, audio, or image that displays in the AR software. Alternately, for an AR video, the tutor might suggest the writer pull up the accompanying video transcript that the tutor and the writer could read aloud together. If the AR displays a static infographic-style image, they might read aloud the text on the image displayed in the AR software. If they notice certain images in particular, the tutor might narrate their reaction to those images and ask questions about citing the sources of the images. Regardless of the AR composition's format or development state, the tutor will talk with the writer to set an agenda for talking through the text.

Talking through the text requires mutual negotiation of *ludus* and *paidia* goals. Because *ludus*-related goals tend to focus on better-written products and *paidia*-related goals focus on better writing, in a game-oriented version of what Stephen North (1984) might say, we contend that tutoring negotiates emergence and progression. Juul (2002) explains the importance of theorizing emergence: "Many games are simple to learn to play, but knowing how to play is not sufficient to play the game well: There is more to playing games than simply memorising the rules. So we need a framework for understanding how something interesting and complex (the actual gameplay) can arise from something simple (the game rules)." Similarly, writing and tutoring are about more than memorizing a list of rules for genres and media or consultation interactions. Juul (2002) goes on to offer a distinction: "Progression games have walkthroughs: lists of actions to perform to complete the game. Emergence games have strategy guides: rules of thumb, general tricks." Those rules of thumb and tricks are less about foreclosing possibilities or being completely free; instead, emergence is a third space somewhere between "completely specifying what *can* happen, and leaving everything to the user/reader/player." A focus on emergence and *paidia* emphasizes creativity, spontaneity, and improvisation.

Agenda in Action

The middle of a session is where the main ludic game play happens as the writer and tutor put the established agenda into action. Fitzgerald and Ianetta (2016) suggest that tutors start with global issues: "the writing as a whole, its ability to communicate with the reader, and its overall effect" (73). Addressing global concerns involves both *ludus* and *paidia.*

A *paidia* focus is mainly at play when the writer vaguely states what they want to work on; for example, wanting the writing to sound right or clear or their AR video or animation content to flow. While tutors are encouraged to dig deeper and ask more clarifying questions to help further pinpoint a clear objective, sometimes the writer does not have the language to fully express their concerns. Therefore, the tutor can respond with exploratory *paidia*-focused questions while they search for the writer's goal. The tutor may ask the writer to explain the main point or central idea of the writing or to walk them through the organizational structure of the project. For AR, the tutor may ask the writer to identify how they want the audience to benefit from the composition's use of the AR medium, to understand how the writer has availed themselves of the affordances of AR. In planning ahead for the marketing AR assignment, the tutor might ask the writer to brainstorm how the design of the video could match the expectations set by the paper trigger.

Global Concerns

Global issues move from *paidia* to a ludic focus when a set goal is defined—for example, if the writer is unable to identify the organization of the composition or wants to make sure their thesis is clear and well supported throughout the project. Elliott Freeman, in his chapter in this volume, explains how he uses a ludic angle to help writers compose a thesis statement through crafting multiple potential statements and negotiating the strengths and weaknesses of each. Before reviewing the AR project, the tutor may ask the writer to state the main purpose and outline the visual, auditory, or textual materials they used to support the purpose. Then, as the writer and tutor look at the AR, they can discuss these elements in detail, looking to see how they connect to the claim and effectively support the purpose. The tutor and writer may also choose to reverse outline the composition, create a storyboard, or otherwise sketch an overview to determine if the current structure is logical. All these steps are intrinsically ludic game play, as they are necessary to build toward the extrinsic goal identified by the writer. In addition, both the writer's intrinsic and extrinsic *ludus* steps build into the tutor's ludic goal of helping the student become a better writer. This work is performed in traditional writing formats as well as those of emerging media. Strategies such as reading aloud, creating a reverse outline, or storyboarding that are employed for responding to global concerns can be reused either for future work on the project or for different upcoming projects. Because these strategies can be redeployed in the future, they carry the re-playability characteristic of emergence.

Local Concerns

Whether located in a lab report or the onscreen text of an AR project, sentence-level local issues tend to have a more ludic focus. The strategies used for identification, such as reading aloud, can include elements of both *paidia* and *ludus* depending on whether the discussion is used as an exploratory tool or to help writers identify patterns. A translingual approach to sentence-level differences focuses on negotiating meaning, which is *paidia*-based as it "opens up a conversation for talking about reading, language choices, and making meaning" (Krall-Lanoue 2013, 230) instead of the tutor assuming an error has been made. Likewise, tutors can approach AR compositions by asking questions about the writer's choices. For example, they might ask the writer why they chose a different font on this text or why this image looks different from another. Some choices may be intentional, and thus it can be valuable to verify the writer's intent. In a marketing AR assignment, the writer might intentionally use slang terminology to catch the audience's attention. Similar to the ways interventions for global issues can be reused for future writing and tutoring, tutors may likewise use these re-playable questions in the future, and writers may be able to use these questions and answers to articulate their writing choices in the future. While games of progression require strategies that are often specific to particular games, the creativity of emergent game play can provide ongoing benefits.

Flexibility

While helping the writer navigate the assignment, the tutor should be flexible within the session that engages in *paidia*-type practices. There is no one prescribed consultation approach; tutors must be flexible and improvise based on the situation. As a result, the tutor should be open to moving back and forth from global and sentence-level local concepts utilizing both *paidia* and *ludus* game play: "Sometimes global and sentence-level issues are so deeply interrelated that it is difficult to separate them. For instance, you might come across a sentence in which a writer has difficulty clearly articulating an idea but that, if revised, could help readers understand the entire argument of the piece" (Fitzgerald and Ianetta 2016, 74). Therefore, flexibility is key.

Although games are bound by rule negotiation, players can employ flexibility within those rules by approaching new games or new scenarios in game play through experimentation, trying previously used strategies, using new combinations, and exploring all the options that may be available to them. In other words, players employ creative problem solving to accomplish game-play actions, just as writers and tutors do, all within the rules circumscribing the activity. As Juul (2002) explains, the concept of emergence "is helpful in understanding how the simple rules in a game can create the variety we often witness." Navigating rules that guide writing and tutoring can lead to improvisation that yields productive creative variations.

WRAPPING UP THE SESSION AND LOOKING FORWARD

The end of a session returns to the negotiation of *ludus* and *paidia* game play. The tutor signifies to the writer that the *paidia* time limit is approaching its end. At this point, the tutor asks the writer if they have any additional questions, reactivating the *paidia* game play. Depending on the writer's response, the tutor may provide the answer, continuing the ludic game play—for example, if the writer requests assistance on citation style for the reference list (just as traditional essays and videos have citations, so do AR projects that incorporate other sources). The tutor may, if time allows, review the reference list and provide resources like the Purdue OWL. The tutor may also encourage the writer to make a follow-up appointment. The tutor has shown the writer how to "play through" the writing and tutoring process, navigating short- and long-term goals, and the writer can leave to continue working through the composing process on their own—negotiating the assignment, genre, and personal rules for the piece.

At this stage, the tutor may also review what was accomplished in the session, walking the writer through the steps taken to accomplish their winning goals. This review is also useful to remind the writer how to approach this game play in the future when the tutor is not present. Here, the tutor assists with planning similar to *Pokémon Go*'s Professor Willow when he appears to offer players guidance for future work but is not otherwise present.

In addition, the tutor will help the writer create a plan for continuing the writing process. The tutor may ask an exploratory *paidia*-style question to initiate a ludic game for the writer by saying, "What do you still need to accomplish as you continue to work on this writing?" As Melanie R. Weaver (2006) states, students may value the feedback they receive, but they do not always know what to do with it. Together, the tutor and student create a plan or similar outline as they did at the beginning of the session to give the writer direction for how to complete their assignment. Fitzgerald and Ianetta (2016) state that "some writers don't know yet how to see their writing processes in terms of doable, discrete steps, and it might not be obvious to them once they leave the session how they should get back to work" (78). Such confusion may occur particularly when writers are working with unfamiliar media and genres. For the marketing AR assignment, the tutor might encourage the writer to come back in after creating a storyboard and script for the digital video or when they're ready for feedback on a rough cut of the video. By walking through future plans with the tutor, writers have a clear objective and purpose for their writing.

While the writer may accomplish and complete the ludic game—for example, by creating an effective thesis, coordinating sound effects with concepts in a logical manner, developing a coordinating AR trigger and digital AR content, or presenting a cohesive visual aesthetic, as set forth as the writer's main goal—the overall goal to become a better writer is still in play. The project is not complete when the session is over, and the writer is not done composing and developing when they submit the assignment. As previously discussed, games of progression focus on the play-through to the conclusion. Juul (2002) notes that games of progression can be identified if a walkthrough guide listing the actions to complete the game can be found online. Emergence is typified by "strategy guides: rules of thumb, general tricks" ("Conclusion"). *Pokémon Go* is a game with no clear end because even if a player could actually "catch 'em all," they can continue to level up their Pokémon to obtain stronger monsters and engage in social play. Emergent complexity in choice, method, and ultimately agency can also be found in both tutoring and writing. By contrast, applying strategies of progression to writing can lead to written products that all look the same. Emergent tutoring fosters development of strategies and guidelines that can be reused beyond the individual assignment or tutoring session.

While we have focused here on thinking about tutoring emerging media such as AR, we find *Pokémon Go* useful as a tool to consider tutoring and writing in an era of change. The mobile game has continually changed since its release in 2016. Further, it is a game of emergence, the opposite of a one-and-done game. For millions of players, *Pokémon Go* has remained highly re-playable as new updates offer new scenarios and activities, with play balanced between performing familiar actions and learning new ones. Much as players negotiate between *ludus* and *paidia* play in games of emergence, tutors and writers similarly navigate these rules and game play, whether in traditional or emerging media. Continually learning to tutor and write in new media and genres involves adopting the emergence mind-set, recognizing that there is no singular, final conclusion or end point in working with writing but that prior familiarity with negotiating the rules of tutoring and writing can be applied again when working with unfamiliar media. Indeed, given that new media continue to emerge, tutors and writers must likewise continue to employ iterative processes of applying familiar processes and learning new strategies.

As this section explores, analyzing tutor interaction using this game-oriented framework to examine strategies, forms of play, and highly replayable games of emergence enables tutors to see the commonalities between tutoring traditional academic writing and composing in unfamiliar media forms and then to apply past strategies to working with new projects. Rather than play-through-once activities, emergent tutoring and writing are ongoing processes with strategies that can be employed beyond a single assignment.

CONCLUSION: TUTORING AR AND BEYOND THROUGH EMERGENT TUTORING

Our discussion in this chapter has focused on conceiving of tutoring through gaming as a strategy for writing center tutoring of AR, an emerging medium that may not yet be a standard part of writing classes, as well as other multimodal projects. As our analysis of tutoring sessions through *ludus* and *paidia* shows, our center's tutors, although they may not always feel prepared to address new forms of texts, do already have foundational preparation to enable them to respond to digital composing. Pemberton (2003) notes that tutors are familiar with "the conventions of print texts, audience, organization, and argument, and they will also likely be skilled users" of digital text (20–21). We similarly contend that tutors are already prepared to be engaged audience members for various subject matters, genres, and media. However, as Pemberton suggests, writing centers can prepare for new media tutoring. Likewise, we propose incorporating activities focused on improvisation and spontaneity for preparing for AR tutoring and composing. For example, many students are familiar with *Pokémon Go*. Whether tutors are experienced or new players, developing emergent tutor training in which tutors play *Pokémon Go* can provide an opportunity for them to discuss how they confront unfamiliar experiences, whether the game itself or new features in the game. Training discussion can focus on how tutors employ flexible and spontaneous strategies to play the game in accordance with its rules. The training session can also engage with looking at AR texts and AR composing software. Rather than provide many specific directions on how to use the AR software, the session can instead encourage tutors to work together and to employ the same experimentation they just used in playing *Pokémon Go.* The sort of improvisation, flexibility, and spontaneity they employed in game play can carry over into their composing and ultimately their tutoring responses.

Such discussion provides the opportunity for reinforcing how game players, tutors, and writers employ *paidia* and *ludus.* Indeed, the advantage of game-inspired tutoring is *paidia*-oriented exploration of the available range of options, which offers a twofold benefit. First, as the introduction to this volume shows, playful exploration fosters creativity, which can be used to resist genre and media expectations that drive end products to look the same as other texts. Second, emergent tutoring fosters developing game-play–informed, problem-solving strategies that writers can employ on their own in the future. Writing centers should strive to help writers approach writing as a highly re-playable activity throughout their futures.

Approaching tutoring and tutoring preparation through gaming reinforces the discovery of options and problem solving. Digital media training can equip tutors to work with specific technologies; however, "it can in no way equip them with all the specialized knowledge they need" (Fishman 2010, 61). As such, training can orient toward the practice of "technology solving so that, regardless of the situation, [tutors] are adept at identifying the problem's origin and locating the proper sources to address it" (61). Whether those challenges are technical or authorial, the ability to use game-play experimentation, collaboration, emergence, and negotiation of rules supports tutors and writers alike in discovering solutions.

Tutoring AR, as with other media, involves talking through a range of multimodal options, including decisions about how to compose; whether to use static, audio, or video digital overlays; what spatial orientation to employ on the display device; whether to use a full range of multimodal options in their AR overlays; and even whether AR is the best medium for their work (Sheridan 2010). By preparing tutors to employ game-play theory to attend to "visual and audio design principles, the rhetorical nature of multimodal composing, and a variety of multimodal genres" (Sabatino 2019, 1), we prepare writers to consider those same future choices when they're composing for AR or other media. Indeed, emergence is a key element in working with writing in an era of changing genres and media, such as AR, by supporting resilience and acquisition of strategies like exploration, invention, and experimentation.

Analyzing session interactions through game play–related concepts provides an approach for supporting tutor response to multiple forms of writing, whether traditional linear alphabetic texts or branching multimodal digital compositions. Game play—*ludus* and *paidia* specifically—helps both tutors and writers approach all forms of writing as a process that requires iterative exploring, improvising, and thinking creatively and critically. Just as players level up in highly replayable emergent games with no clear termination, a tutoring theory of emergence recognizes that a written product isn't the ultimate objective but instead that ongoing encounters with writing will yield new challenges, new opportunities for exploration, creativity, and discovery in new media. Through a tutoring theory of emergence, writing centers can support the processes that foster the flexibility and spontaneity required for tutoring and writing across all media: traditional, emerging, or yet to exist.

NOTE

1. To access an AR text, the audience must activate the AR display software on their smartphone and then pointing the phone's camera at a visual image. Some AR composers may use existing, physical triggers, such as statues or building signs, as the triggers for their AR texts. As noted earlier, the AR author might be working on a project to augment campus buildings with digital AR that explains the history of individual buildings or, perhaps, the architectural changes across time. A tutor might note how they feel as they attempt to trigger the AR. They might identify whether they think the use of a building versus its sign is clear or confusing to use as a trigger. They might ask the AR author how a choice to use the image of the building itself versus the building's sign as a trigger impacts the audience; for example, the tutor might ask which is easier for the audience to use as a trigger: the building itself or the building sign. The tutor might ask the AR author which is consistent with other AR texts within the larger AR project discussing campus architecture. For other projects, AR composers might choose to design a visual trigger, such as a paper poster in which the AR text offers digital media that augment the static content on the poster. Tutoring such a trigger might involve asking the writer what information should be located on the printed poster and what information might be overwhelming or distracting to the audience. The tutor might inquire about various design choices, such as graphics and color usage, or how much information should be presented on the printed poster and how much should be delivered in the digital text and how. In other words, AR triggers are as much a part of the AR tutoring conversation as the AR digital composition.

REFERENCES

Alberti, John. 2008. "The Game of Reading and Writing: How Video Games Reframe Our Understanding of Literacy." *Computers and Composition* 25, no. 3: 258–69.

Ang, Chee Siang. 2006. "Rules, Gameplay, and Narratives in Video Games." *Simulation and Gaming* 37, no. 3: 306–25.

Arth, Clemens, Raphael Grasset, Lukas Gruber, Tobias Langlotz, Alessandro Mulloni, and Daniel Wagner. 2015. *The History of Mobile Augmented Reality. arXiv.* https://arxiv.org/pdf/1505.01319.pdf.

Balester, Valerie, Nancy Grimm, Jackie Grutsch McKinney, Sohui Lee, David M. Sheridan, and Naomi Silver. 2012. "The Idea of a Multiliteracy Center: Six Responses." *Praxis: A Writing Center Journal* 9, no. 2. http://www.praxisuwc.com/baletser-et-al-92/.

Blevins, Brenta. 2018. "Teaching Digital Literacy Composing Concepts: Focusing on the Layers of Augmented Reality in an Era of Changing Technology." *Computers and Composition* 50: 21–38.

Caillois, Roger. 1961. *Man, Play, and Games.* Translated by Meyer Barash. New York: Free Press of Glencoe.

Carpenter, Rusty, and Sohui Lee. 2016. "Introduction to the Special Issue: Envisioning Future Pedagogies of Multiliteracy Centers." *Computers and Composition* 41: v–x.

Colby, Rebekah Shultz. 2017. "Game-Based Pedagogy in the Writing Classroom." *Computers and Composition* 43: 55–72.

Colby, Rebekah Shultz, and Richard Colby. 2008. "A Pedagogy of Play: Integrating Computer Games into the Writing Classroom." *Computers and Composition* 25: 300–312.

Delello, Julie A., Rochell R. McWhorter, and Kerri M. Camp. 2015. "Integrating Augmented Reality in Higher Education: A Multidisciplinary Study of Student Perceptions." *Journal of Educational Multimedia and Hypermedia* 24, no. 3: 209–33.

Fishman, Teddi. 2010. "When It Isn't Even on the Page: Peer Consulting in Multimedia Environments." In *Multiliteracy Centers: Writing Center Work, New Media, and Multimodal Rhetoric,* edited by David M. Sheridan and James A. Inman, 59–73. Cresskill, NJ: Hampton.

Fitzgerald, Lauren, and Melissa Ianetta. 2016. *The Oxford Guide for Writing Tutors: Practice and Research.* New York: Oxford University Press.

Förster, Lisa-Katharina, and Folker Metzger. 2015. " 'Time Window Weimar': Students Map Their Town's History through Augmented Reality." *Hyperrhiz* 12. http://hyperrhiz.io/hyperrhiz12/augmented-maps/2-forster-metzger.html.

Frasca, Gonzalo. 1999. "Ludology Meets Narratology: Similitude and Differences between (Video)Games and Narrative." *Ludology.org.* http://www.ludology.org/articles/ludology.htm.

Gee, James Paul. 2003. *What Video Games Have to Teach Us about Learning and Literacy.* New York: Palgrave Macmillan.

Greene, Jacob W. 2018. "Creating Mobile Augmented Reality Experiences in Unity." *Programming Historian.* Last modified June 23, 2021. Retrieved from https://programminghistorian.org/en/lessons/creating-mobile-augmented-reality-experiences-in-unity.

Grouling, Jennifer, and Jackie Grutsch McKinney. 2016. "Taking Stock: Multimodality in Writing Center Users' Texts." *Computers and Composition* 41: 56–67.

Juul, Jesper. 2002. "The Open and the Closed: Games of Emergence and Games of Progression." In *Computer Games and Digital Cultures Conference Proceedings,* edited by Frans Mäyrä, 323–29. Tampere, Finland: Tampere University Press. http://www.jesperjuul.net/text/openandtheclosed.html.

Juul, Jesper. 2005. *Half-Real: Video Games between Real Rules and Fictional Worlds.* Cambridge, MA: MIT Press.

Krall-Lanoue, Aimee. 2013. " 'And Yea I'm Venting, but Hey I'm Writing Isn't I': A Translingual Approach to Error in a Multilingual Context." In *Writing as Translingual Practice in Academic Contexts,* edited by Suresh Canagarajah, 228–34. London: Routledge.

Lochman, Daniel T. 1986. "Play and Game: Implications for the Writing Center." *Writing Center Journal* 7, no. 1: 11–18.

Macauley, William. 2005. "Setting the Agenda for the Next Thirty Minutes." In *A Tutor's Guide: Helping Writers One to One,* edited by Ben A. Rafoth, 1–8. Portsmouth, NH: Boynton/Cook.

McKinney, Jackie Grutsch. 2009. "New Media Matters: Tutoring in the Late Age of Print." *Writing Center Journal* 29, no. 2: 28–51.

Morris, Jill. 2019. "Augmented Reality Design through Experience Architecture." In *Not Just Play: Essays on Motivations and Impacts of Pokémon Go*, edited by Jamie Henthorn, Andrew Kulak, Kristopher Purzycki, and Stephanie Vie, 62–78. Jefferson, NC: McFarland.

Niantic. 2020. "*Pokémon Go.*" *Google Play*, 0.169.0-G. https://play.google.com/store/apps/details?id=com.nianticlabs.pokemongo&hl=en_US.

North, Stephen. 1984. "The Idea of a Writing Center." *College English* 47, no. 5: 433–46.

Pemberton, Michael. 2003. "Planning for Hypertexts in the Writing Center . . . or Not." *Writing Center Journal* 24, no. 1: 9–24.

Rouzie, Albert. 2000. "Beyond the Dialectic of Work and Play: A Serio-Ludic Rhetoric for Composition Studies." *Journal of Advanced Composition* 20, no. 3: 627–58.

Sabatino, Lindsay. 2009. "Comfort Empowers." *Praxis: A Writing Center Journal* 6, no. 2. http://www.praxisuwc.com/sabatino-62.

Sabatino, Lindsay. 2014. "Improving Writing Literacies through Digital Gaming Literacies: Facebook Gaming in the Composition Classroom." *Computers and Composition* 32: 41–53.

Sabatino, Lindsay. 2019. "Introduction: Design Theory and Multimodal Consulting." In *Multimodal Composing: Strategies for Twenty-First-Century Writing Consultations*, edited by Lindsay A. Sabatino and Brian Fallon, 1–16. Logan: Utah State University Press.

Sheridan, David M. 2010. "All Things to All People: Multiliteracy Consulting and the Materiality of Rhetoric." In *Multiliteracy Centers: Writing Center Work, New Media, and Multimodal Rhetoric*, edited by David M. Sheridan and James A. Inman, 75–107. Cresskill, NJ: Hampton.

Tham, Jason, Megan McGrath, Ann Hill Duin, and Joseph Moses. 2016. "Glass in Class: Writing with Google Glass." *Journal of Interactive Technology and Pedagogy* [Assignments section]. http://jitp.commons.gc.cuny.edu/glass-in-class-writing-with-google-glass.

Tosca, Susana. 2003. "The Quest Problem in Computer Games." In *Technologies for Interactive Digital Storytelling and Entertainment (TIDSE) Conference*, edited by Stefan Göbel, Rainer Malkewitz, and Ido Iurgel, 69–82. Darmstadt: Fraunhofer IRB Verlag.

Trimbur, John. 2000. "Multiliteracies, Social Futures, and Writing Centers." *Writing Center Journal* 20, no. 2: 29–31.

Weaver, Melanie R. 2006. "Do Students Value Feedback? Student Perceptions of Tutors' Written Responses." *Assessment and Evaluation in Higher Education* 31, no. 3: 379–94.

Yancey, Kathleen Blake. 2004. "Made Not Only in Words: Composition in a New Key." *College Composition and Communication* 56, no. 2: 297–328.

6

THE WRITING CONSULTATION AS FANTASY ROLE-PLAYING GAME

Christopher LeCluyse

Central to the social turn that has marked writing studies for the past three decades is the notion of identity as performance. Nancy Grimm (1999) sparked discussion of this concept in the writing center community with her book *Good Intentions: Writing Center Work for Postmodern Times*, which discusses writing centers' sometimes fraught role in helping student writers perform what Harry Denny (2010) later called "the codes of cultural dominance" by reinforcing the norms of academic writing to the exclusion of underrepresented students (75). Denny's (2010) foundational book *Facing the Center: Toward an Identity Politics of One-to-One Mentoring* consistently examines various forms of identity—race and ethnicity, class, sex and gender, and nationality—through the terms *perform, performance*, and *performativity* no fewer than sixty-five times. When writing center scholars do take up the performative nature of identity, however, their focus has generally been on consultants or on the identity of the writing center itself rather than on writers (see, for example, Gladstein 2007; Griffin et al. 2006). Denny's work notwithstanding, there is little discussion of *how* student writers construct academic identities in the writing center or what roles consultants play in modeling or helping them construct those identities. Likewise, little work examines how writers come to understand the writing consultation as a system—the rules of the consultation itself, not just of academic writing.

To help address this need, I synthesize two perspectives on identity formation: threshold concepts of writing and of writing centers (Adler-Kassner and Wardle 2015) and the social scientific study of fantasy role-playing games (RPGs) such as *Dungeons & Dragons*. Key to threshold concepts of writing is the notion that written identities are performed—in other words, a form of role playing. The social scientific examination of role-playing games pioneered by Gary Alan Fine (1983) provides a conceptual framework for understanding how student writers negotiate

https://doi.org/10.7330/9781646421947.c006

the various frames of reference they are required to inhabit as they perform a written persona. Understanding the performances writers act out in the writing center as a form of fantasy role playing exposes the choices they must make to address an academic discourse community. Seeing the consultation as a role-playing game also reveals the conventions of writing center praxis as their own bounded system and helps us understand how writers come to learn those conventions through experience. Writing consultants best serve writers in this game by making the moves of both systems explicit, using meta-commentary to demystify not only academic writing but also the writing consultation itself. By doing so, rather than forcing student writers on an over-determined hero's quest, writing center practitioners can create a playful space to explore alternative subjectivities.

ACADEMIC WRITING AS ROLE PLAYING

Before examining how negotiating identity in fantasy RPGs helps us understand the choices writers make in the writing center, we should acknowledge that producing academic writing requires writers to role play. Indeed, the understanding that writing is a performance of identity is one of the threshold concepts of writing studies. Linda Adler-Kassner and Elizabeth Wardle (2015) define threshold concepts as "concepts critical for continued learning and participation in a community of practice" (2). Among these threshold concepts is the notion that "writing enacts and creates identities and ideologies" (Scott 2015, 48). As Kevin Roozen (2015) explains, "Through writing, writers come to develop and perform identities in relation to the interests, beliefs and values of the communities they engage with, understanding the *possibilities of selfhood* available in those communities . . . The act of writing, then, is not so much about using a particular set of skills as it is about *becoming a particular kind of person*" (50–51, emphasis added). Fundamental to our current conception of writing, then, is that each act of writing involves performing a persona. This is especially the case when writing for academic audiences, as academic readers frequently have robust and fixed notions of the identities they expect writers to perform—to the reward of those who perform accordingly and to the detriment of those who don't.

Considered in isolation, "performing identities," "understanding possibilities of selfhood," and "becoming a particular kind of person" could very well define the key focus of role-playing games. Central to role playing is the performance of a character, which requires imagining an alternate identity. True to a postmodern notion of identity, this threshold concept presents the self—or particular versions of a self—as not merely expressed but as *constructed* through writing, just as an RPG character is constructed through game play.[1] As David Schaafsma (1998) explains, citing Michel Foucault, "identity categories . . . must be seen as historical events that are continuously performed" rather than essential qualities "that are just waiting to be expressed" (264). So, too, a player's character in an RPG is not merely a projection of their daily persona—indeed, it cannot be, since that character operates in a fantasy world fundamentally different from our own—but is constructed through the act of playing. Events in the game, even individual dice rolls, continually shape how the player performs that persona, and players often intentionally choose to construct characters whose personalities and moralities are quite different from their own.

These identities are not performed in isolation, however. In both academic writing and the RPG, the performance takes place within a community—be it the larger community of higher education, a particular discipline, or all players of a particular RPG or within the intimate community of what Fine (1983) terms an idioculture, "a system of knowledge, beliefs, behaviors, and customs peculiar to an interacting group" (136). While readers generally expect writers to display identities in line with the values of the communities for and in which they are writing, writers may choose to "challenge, perhaps even contest and resist, [their] alignment with the beliefs, interests, and values" of those communities (Roozen 2015, 51). For example, college faculty generally expect students to write in Standard American English, but students who speak undervalued dialects or other languages may choose to code mesh by using the other language varieties at their command in their academic writing—an approach Neisha-Anne S. Green (2016) advocates.

Even in situations where writers prefer to perform expected identities, they may experience challenges because they have difficulty "see[ing] themselves as participants in a particular community" (Roozen 2015, 51). For three decades, writing scholars have traced the difficulties student writers often experience when trying to perform for academic audiences. For example, in his much-cited article "Inventing the University," David Bartholomae (1986) presents the plight of basic writers who "have to appropriate (or be appropriated by) a specialized discourse, and they have to do this as though they were easily and comfortably one with their audience, as though they were members of the academy, or historians or anthropologists or economists . . . They must learn to speak our language. Or they must dare to speak it, or to carry off the bluff, since speaking and writing will most certainly be required long before the skill is 'learned' " (5). In other words, students must perform confident, competent academic selves long before they have mastered the moves of academic writing.

Often, the identities students are expected to perform reflect middle-class, white values. Donna LeCourt (2006) addresses the way working-class students frequently come up against differences between "the classed nature of experience coded in [their own] discursive strategies" and "those we teach in first-year composition courses" (30). In this environment, students feel pressured to perform a middle-class identity and to jettison a working-class one. Finally, students of color likewise frequently feel such disjunctions between the identities they are expected to perform as academic writers, even within the supposedly welcoming confines of the writing center, and those they have cultivated at home. Green (2016) expresses her experience as "an international graduate student of color" as such: "There seemed to be no vocabulary that I was aware of within any of the discourse communities that I 'belonged' to, especially ours in the writing center, that adequately summed up and fully described what I felt deep within my being. Because of this, I had not been able to tell anyone that I needed help, or that I was struggling with any one particular thing. No one knew that I fought for years at school to be my best academic self and still maintain a semblance of my identity" (72). Notice the distinction between Green's "best academic self" and what she sees as her "identity." Succeeding in higher education required her to role play, in her case by excising any traces of African American or Barbadian English from her writing.

Although most discussions of the complexities of negotiating identity in an academic context take place outside the framework of threshold concepts, they reflect the threshold concept that "disciplinary and professional identities are constructed through writing" (Estrem 2015, 55). As Heidi Estrem (2015) explains, "The process of learning to manage these tensions [between the identities writers are expected to perform and how they see themselves] contributes to the formation of new identities" (56). Student writers have no way out but through, in other words, and they develop as writers precisely by negotiating new challenges and performing new written identities to meet those challenges (see Carroll 2002, 9). The writing center can therefore provide the scaffolding to identify and respond to such challenges.

Writing centers serve as sites where student writers practice constructing identities that may occupy a range of positions with respect to those expected by readers in particular communities. As Rebecca

S. Nowacek and Bradley Hughes (2015) observe, writing centers are "hardly neutral sites of literacy instruction or assistance" (171). While some aim to conform students to the values of academic discourse communities, other seek to liberate them from those norms. Such writing centers, which Melissa Ianetta (2004) terms "Socratic" in that like Socrates they "sp[eak] truth to power and embrace [their] outsider role, . . . embrace the margins as a countercultural rhetorical space from which to critique and improve disciplinary and institutional hierarchies" (46). The choices writing consultants make therefore have significant stakes attached. They can encourage writers to make choices that align their written identities with academic customs and power structures—a choice that may garner higher grades and other tokens of success—or they can open up alternative identities that may buck academic convention but come closer to how writers actually see and express themselves.

The study of fantasy RPGs yields a powerful framework for understanding how writer-players situate themselves with respect to the identities they perform and the system through which they perform them. It also delineates the role of the writer-player from that of the writing consultant, who could be seen as a Dungeon Master. These distinct and asymmetrical roles are common to most role-playing games. Players control only their own characters' actions in the fantasy world while a referee-narrator, called a Dungeon Master in *Dungeons & Dragons*, sets up the story those characters are a part of, controls the fantasy world and its other inhabitants, and adjudicates the rules.

While the performance of identity and the interaction of player and referee may be true for any genre of RPG (e.g., superhero, science fiction, cyberpunk, cartoon), the nature of the self makes what happens in academic writing and in the writing center forms of *fantasy* role playing. Ira J. Allen (2018) argues that any notions of the self and of agency are in fact fantasies—though necessary ones. They are necessary because for us as social and symbolic beings, "self and agency are instruments of negotiation" (176). We need to perform particular selves to negotiate the constraints of reality "in order to act differently in the world" (177). The work of writing center practitioners is to help students in "negotiating *good* fantastical selves and with discovering *good* ways of accomplishing such negotiation" (189, original emphasis).[2] Such "good" selves, negotiated well, help student writers achieve their personal and academic aims without sacrificing important aspects of their identities or sacrificing their identities to that of a monolithic, idealized academic writer.

FRAMES OF REFERENCE IN THE RPG AND THE WRITING CONSULTATION

The growing study of social interactions in RPGs provides writing center practitioners with detailed understandings of how individuals generate fantastical selves. In the most popular role-playing game, *Dungeons & Dragons,* players embody characters in a medieval fantasy setting, typically undertaking some kind of quest in a world constructed and mediated by a Dungeon Master. The rules of the game and the rolling of polyhedral dice add structure to this exercise in improvised performance and collaborative storytelling by setting parameters regarding what characters can and cannot do and by adding an element of chance. Seen in this light, the writing consultation becomes an RPG in which a consultant (Dungeon Master) collaborates with a writer (player) to co-create both the discursive world the writer wishes to inhabit and the writer's persona in that world (see Thompson et al. 2009, 97). While treating the consultation as a form of performance is not new, past treatments of the topic (Hemmeter 1994) focus on how consultants construct identities and negotiate differences in power. My interest here is to expose the psychological and social dynamics of how writers construct identities in the writing center and how consultants help writers learn and play the dual games of academic writing and the writing consultation. To that end, I will first explain the system of frame analysis developed by Fine before relating it to students' performance of identity in their writing and in the writing center.

Fine (1983) published *Shared Fantasy: Role-Playing Games as Social Worlds* less than a decade after the initial publication of *Dungeons & Dragons.* His work is therefore striking both for its rigor and its proximity to the birth of fantasy role playing. Fine begins with the social performance theory of Erving Goffman, whose work also inspired Thomas Newkirk's (1995, 1997, 2004) studies of academic writing as performance. Thomas "Buddy" Shay and Heather Shay's chapter in this volume likewise uses social performance theory to understand tutor identities as fantasy role playing. Goffman himself applies his system of frame analysis to social interactions in games, laying out how "our understanding of the rules of framing and of organizing game experiences are [*sic*] acquired in the 'outside' world, and are [*sic*] required for the structuring of a play world" (quoted in Fine 1983, 183). In this one pronouncement, Goffman identifies the three frames of reference or "levels of meaning" Fine (1983) employs in his study (186). The "outside world" corresponds to what Fine (1983) calls "the 'primary framework,'

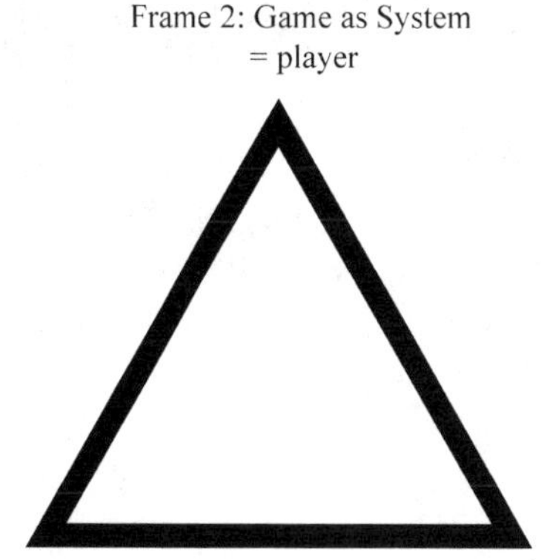

Figure 6.1. Frames of reference in fantasy RPGs

the commonsense understandings that people have of the real world" (186). This framework considers the participant in an RPG as a *person*. In the second frame, the participant is regarded as a *player*. Goffman's "rules of framing and of organizing game experiences" depend heavily on the conventions and mechanics of the game as a system. Finally, Goffman's "play world" equates with Fine's third frame of reference, the imagined world created through the game experience. Fine (1983) refers to this imagined projection of the participant as their character (186), or we may for the sake of alliteration follow Dennis Waskul (2006) in calling this aspect the persona (26). Shay and Shay similarly utilize Waskul's tripartite division in their chapter.

Figure 6.1 represents the three frames of reference. Labels juxtapose Fine's and Waskul's terminology.

Distinguishing among person, player, and persona enables us to analyze participants' individual approaches to RPGs, their experiences when playing, and difficulties they may encounter.[3] A challenge of role playing is distinguishing between what the person in the present-day real world knows and what their persona in the medieval fantasy setting knows. As Fine (1983) explains, players often make decisions for their characters based on real-world knowledge their fantasy personas would not be expected to have (188). Key to the norms of RPG players, however, is maintaining a pretense that such is not the case. In this "pretense awareness context," participants partition their various selves and try to behave as if each was unaware of the other (188). When this norm is ruptured, however, players may level accusations of meta-gaming, having one's persona act based on personal out-of-game knowledge in a way that gives them an unfair advantage or at least violates the simulation of a fantasy world (Bowman 2010, 114–15; Laycock 2015, 11).

Occasionally, confusion results from the opposite problem: situations in which an in-game persona knows something the player does not know. As Fine (1983) explains, "The character is limited by the insight of his player" (209). Joseph P. Laycock (2015) illustrates this point with a story that has passed into urban legend within the fantasy RPG community (11–12). A teenage player who asked the Dungeon Master what his character saw upon stepping into a clearing was told "you see a gazebo." The player, not familiar with the term and thinking that a gazebo was some kind of otherworldly creature, then proceeded to try to observe and address the gazebo and, fearing a threat, finally shot an arrow into it, much to the Dungeon Master's and his fellow players' amusement. In the game world, of course, no such confusion would have occurred: the character would have seen a structure he ostensibly understood and not treated it like a threat. A similar problem can arise when the person lacks knowledge of the game as a system. They may have an idea of what they want their persona to achieve but do not know how to do so within the mechanics of the game.

Such overlap and occasional disjunction likewise can occur as academic writers construct their personas as writers. While RPG players may orient themselves toward their characters in a variety of ways, academic writers are more like players who in Fine's (1983) terms "pla[y] a character as an *extension* of one's person—one's person in a more extreme fashion" (208, original emphasis). This orientation may indeed be the most common, for as Waskul (2006) explains, "Participants in role-playing games often find it difficult to play a fantasy persona that is *purely* fictional . . . In fact, many role-players claim that effective play presumes gamers who identify with and otherwise apprehend their fantasy personas as extensions of themselves" (31, original emphasis). Whereas the RPG player can appeal to an additional level of knowledge and experience their character possesses, however, the writer's academic persona is confined to their own knowledge base. Student writers in particular must frequently adopt a different pretense—not that their personal identity is distinct from their academic persona, as would be the case in an RPG, but that their academic persona in fact possesses a level of knowledge and comfort appropriate to experts in the field.

We therefore witness in the writing center how the written performance of identity is a liminal threshold concept, one that "require[s] not just cognitive understanding but sometimes profound reorientations of worldviews and even identities" (Nowacek and Hughes 2015, 175). Students in the writing center can come to understand that the academic selves they perform in their writing are just that—performances distinct from their other selves. With that realization comes a kind of power: in Bartholomae's (1986) terms, they can thereby "appropriate . . . a specialized discourse" rather than "be appropriated by" it (4). Even if they choose to play a character completely in line with the expectations of academic readers, they can do so completely in control of that character rather than unconsciously succumbing to a set of half-articulated expectations. Tutors can assist in this process by making the performance explicit, addressing the performed identity in writing as the constructed character it is.

Recognizing the distinction between person and persona raises the possibility of resisting rather than simply adopting expected identities as academic writers. Sarah Lynne Bowman (2010) identifies nine distinct relationships between person and persona (155–56). These relationships lie on a continuum, ranging from players who treat their characters as mirror images of themselves to those who experiment with alternative, even transgressive moralities.[4] While some players may have their characters act and behave more or less as they would in daily life (albeit with some added heroics and supernatural abilities), others may intentionally make their characters entirely different from themselves, even to the point of being outright evil.

ALTERNATIVE IDENTITIES FOR THE CONSULTANT/DUNGEON MASTER

Treating the writing consultation as a form of fantasy role playing therefore opens up possibilities for consultants beyond that of authority or gatekeeper—alternatives Nowacek and Hughes (2015) include in their "expansive vision of the writing center" (172–74). Conscious of their role as Dungeon Master (DM), they can situate themselves more as enablers of others' fictions, helping them craft the personas they want to perform rather than imposing a master narrative on who they are expected to be. Seeing this range of possibilities creates a playful space for writers to fashion characters who may differ both from writers' past identities and from the norms of some academic discourse communities. Like experienced role players, they may experiment with new selves in their academic writing instead of simply treating their characters as extensions of themselves. To return to my previous example of how writers may choose to resist academic norms, a student may decide to mesh codes and integrate other dialects or languages into their writing, take their writing in a direction *other* than that suggested by faculty, or at the very least use word choices and sentence structures that better capture the voice they want to project rather than the one they feel expected to use. Of course, as in any good role-playing game, such choices come with risks, but supportive consultant/DMs can help writers take on such risks mindfully and in so doing grant them greater agency over their writing.

While the game of performing writing identities is not an easy one, in a writing center context, players do not play that game alone. Many student writers adopt the second, systemic frame of academic writing through trial and error with the occasional input of faculty willing to demystify its moves. Upon entering the writing center, however, they enter a situation even closer to that of an RPG. There is, after all, another participant in the consultation, which like an RPG is an improvised and collaborative experience whose primary purpose is to generate if not narrative, then discourse. True to the asymmetrical setup of RPGs, that participant, the consultant, has a different role as referee or mediator. Like a Dungeon Master, their responsibility is to help construct the imagined world the writer/player wishes to inhabit through their fabricated persona—in this case, the world of academic discourse. The involvement of the consultant also addresses the gazebo scenario in which the writer lacks key knowledge their persona needs to operate successfully in the imagined world. In such situations, the writer can leverage the consultant/DM's knowledge to perform more successfully in the game environment. Just as a DM might explain to a player aspects of surviving in a medieval fantasy world they as real-world individuals may not know (should you drink the water?), consultants can pull back the curtain on academic writing conventions by making them explicit. The second frame is paramount, however: while the consultant cannot "be" the character for the writer, they can model the moves of a rhetorically adept academic writer, thus helping the writer/player craft their own persona. By focusing on academic writing as a system, the consultant can provide a knowledge base for the writer, who can then build on that base to develop their own written identity. While the writer can follow the consultant's lead by taking into account the typical moves of academic writers, they can still make their own choices regarding whether and how to make those moves.

LEARNING THE RULES OF THE WRITING CONSULTATION

Complicating the consultant's explanation of academic writing as a system is the fact that the consultation itself operates as a game. There are in effect two second frames of reference—not only the conventions of academic writing but also the conventions of the writing consultation. Writers new to the writing center must therefore acquire an understanding of how the consultation operates out of game even as they are developing facility with academic discourse in-game.[5] Very little scholarship has examined *how* student writers come to understand the conventions of writing center praxis. Most treatments of the topic instead focus on what are framed as students' misconceptions of writing center praxis or on differences between students' expectations and writing center lore.[6] Applying Fine's frame analysis can help capture the way writers develop a sense of writing center praxis by isolating that praxis as a rule-bound system. (Also see Jamie Henthorn's chapter, in this volume, on using RPG elements to help tutors learn about writing center praxis.) Future observational studies could also track how writers' and consultants' attentions shift among the concerns of the real people interacting in a writing consultation (the first frame), the conventions or academic genres of a writing consultation (the second frame), and the persona constructed in a piece of academic writing (the third frame).

While many writing center scholars present the writing consultation as transformative (see, for example, Abascal-Hildebrand 1994; Geller et al. 2006; LeCluyse 2009), a consideration of the findings of RPG scholars can also provide a mechanism for *how* writing consultations transform writers. Laycock (2015) and Bowman (2010) characterize role playing as a form of ritual, which can create a liminal space that allows participants to scrutinize ordinary reality from an outside perspective. Drawing on Kenneth Burke, Laycock (2015) identifies the central action of this process as "to beyond . . . wherein achieving a higher perspective causes one to think about and interpret the world differently" (74). Effective writing consultations have a similar potential "to beyond" academic writing by providing writers with a liminal space from which they can constructively critique it. Acknowledging the writer's academic voice as a consciously constructed persona is an important aspect of that beyonding.

This awareness also gives writing consultants the opportunity to beyond their own assumptions and practices. Often with the best intentions, consultants may conform writers to a fixed set of expectations, guiding them toward playing the persona that fits the consultant's own notion of academic success. The threshold concept that identities in writing are always performed and multiple cautions against such norming, however. Playing the game of the writing consultation as a DM means leaving writers free to role play their fantastic selves as they wish.

We wouldn't expect an RPG player to play the same character in every adventure: each collaboratively generated narrative calls for a fresh character to suit a new situation. We shouldn't expect writers to do so, either. Instead, the attentive consultant/DM can in effect ask, Who do you want to be today?

Having considered how academic writing and the writing consultation function as fantasy RPGs, I conclude by offering what I think is the best way to act on this knowledge: the use of meta-commentary. By laying out the steps of the consultation up front, responding as a reader, and making the moves of academic writing explicit, consultant DMs can create the space for writers to perform productive written identities. Writers can begin to learn the rules of the consultation game if consultants start the consultation with a preview of what's going to happen—for example, "First, I'll read through your paper and identify the kinds of issues you said you were concerned about. Then, once I'm done reading, we can talk about it and look at ways of addressing those concerns." While reading, consultants can pause to register their responses as readers, thereby helping writers see that they are indeed performing an identity for an audience (Mendelsohn 2012, 82). This form of meta-commentary can involve simply stating how the writer is coming across—"Because you've presented so much evidence, I really feel like you know what you're talking about"—or forecasting where the consultant thinks the writer is going next in the so-called point-predict method (Block 2016, 36–37). Within the consultation proper, such meta-commentary can take the form of "transfer talk" (Hill 2016, 85), which involves explicitly naming the threshold concepts writers are engaging and inviting them to connect those concepts to prior experiences. As Heather N. Hill (2016) explains, such transfer talk can be implicit—simply naming the genre or writing issue in the course of talking about it—or explicit: "I see you're writing a research paper. When's the last time you wrote one? What kinds of issues came up for you then?" Engaging in this form of meta-commentary promotes knowledge transfer by involving students in identifying relevant concepts and abilities cultivated in past learning experiences before extending them to the present task.

All of these strategies go beyond what students are writing to help them think about how they are doing so—to see themselves as players making choices to perform particular identities with the dual systems of academic writing and the writing consultation. Such meta-commentary also foregrounds the threshold concepts Nowacek and Hughes (2015) present as key to overcoming differences in the ways writers and consultants view the consultation. By overtly stating the rules of the game, registering the identities writers are performing, and connecting writing to prior performances, consultants empower writers to make choices. In the process, writers become consciously engaged in a dynamic gaming experience.

NOTES

1. Postmodernist philosophers such as Michel Foucault challenged the notion that identity was inherent or essential, instead insisting that identity is performed. For a more recent treatment of identity as performance, see Allen (2018).
2. While viewing writing as a form of role playing may appear to be particularly (post)modern, it is in fact quite ancient. The classical Greek curriculum in rhetoric and prose writing known as the *progymnasmata* ("preliminary exercises") featured role playing under the guise of *ethopoeia* ("characterization") (Kennedy 2003, xiii), remaining a common feature of European rhetorical education well into the Renaissance (Fleming 2016, 129; Kennedy 2003, ix; Loveridge 2016, 75–76). Ethopoeia involved young rhetors-in-training in creating a speech in the character of a legendary or historical figure. Scholars of ethopoeia note its game-like aspect for both ancient and modern students (Fleming 2016, 131) and identify it as a form of "role-playing" (Walker 2011, 114).
3. Building on and adapting Fine's framework, Daniel Mackay (2001) analyzes RPGs as performance art, and Jennifer Grouling Cover (2010) analyzes them as narrative.
4. Bowman (2010) identifies nine distinct relationships between person and persona as expressed in interviews she conducted with RPG players (155–56).
5. For other treatments of writing center praxis as a form of play, see Dvorak and Bruce (2008); Lochman (1986).
6. See, for example, Harris (1997). Thompson et al. (2009) correlate student and consultant satisfaction with their perceptions of the consultants' directiveness and expertise. For an examination of student conceptions of the writing center in a two-year-college context, see Missakian et al. (2016).

REFERENCES

Abascal-Hildebrand, Mary. 1994. "Tutor and Student Relations: Applying Gadamer's Notions of Translation." In *Intersections: Theory-Practice in the Writing Center*, edited by Joan A. Mullin and Ray Wallace, 172–83. Urbana: National Council of Teachers of English.

Adler-Kassner, Linda, and Elizabeth Wardle. 2015. *Naming What We Know: Threshold Concepts for Writing Studies.* Logan: Utah State University Press.

Allen, Ira J. 2018. "Composition Is the Ethical Negotiation of Fantastical Selves." *College Composition and Communication* 70, no. 2: 169–94.

Bartholomae, David. 1986. "Inventing the University." *Journal of Basic Writing* 5, no. 1: 4–23.

Block, Rebecca. 2016. "Disruptive Design: An Empirical Study of Reading Aloud in the Writing Center." *Writing Center Journal* 35, no. 2: 35–59.

Bowman, Sarah Lynne. 2010. *The Functions of Role-Playing Games: How Participants Create Community, Solve Problems, and Explore Identity.* Jefferson, NC: McFarland.

Carroll, Lee Ann. 2002. *Rehearsing New Roles: How College Students Develop as Writers.* Studies in Writing and Rhetoric. Carbondale: Southern Illinois University Press.

Cover, Jennifer Grouling. 2010. *The Creation of Narrative in Tabletop Role-Playing Games.* Jefferson, NC: McFarland.

Denny, Harry. 2010. *Facing the Center: Toward an Identity Politics of One-to-One Mentoring.* Logan: Utah State University Press.

Dvorak, Kevin, and Shanti Bruce, eds. 2008. *Creative Approaches to Writing Center Work.* Cresskill, NJ: Hampton.

Estrem, Heidi. 2015. "Disciplinary and Professional Identities Are Constructed through Writing." In *Naming What We Know: Threshold Concepts of Writing Studies,* edited by Linda Adler-Kassner and Elizabeth Wardle, 55–57. Logan: Utah State University Press.

Fine, Gary Alan. 1983. *Shared Fantasy: Role-Playing Games as Social Worlds.* Chicago: University of Chicago Press.

Fleming, David. 2016. "Quintilian, Progymnasmata, and Rhetorical Education Today." *Advances in the History of Rhetoric* 19, no. 12: 124–41.

Geller, Anne Ellen, Michele Eodice, Frankie Condon, Meg Carroll, and Elizabeth Boquet. 2006. *The Everyday Writing Center: A Community of Practice.* Logan: Utah State University Press.

Gladstein, Jill. 2007. "Quietly Recreating an Identity for a Writing Center." In *Marginal Words, Marginal Work? Tutoring the Academy in the Work of Writing Centers,* edited by William J. Macauley and Nicholas Mauriello, 211–43. Cresskill, NJ: Hampton.

Green, Neisha-Anne S. 2016. "The Re-Education of Neisha-Anne S Green: A Close Look at the Damaging Effect of 'a Standard Approach,' the Benefits of Codemeshing, and the Role Allies Play in This Work." *Praxis: A Writing Center Journal* 14, no. 1: 72–82.

Griffin, Jo Ann, Daniel Keller, Iswari P. Pandey, Anne-Marie Pedersen, and Carolyn Skinner. 2006. "Local Practices, National Consequences: Surveying and (Re)constructing Writing Center Identities." *Writing Center Journal* 26, no. 2: 3–21.

Grimm, Nancy Maloney. 1999. *Good Intentions: Writing Center Work for Postmodern Times.* Portsmouth, NH: Heinemann.

Harris, Muriel. 1997. "Cultural Conflicts in the Writing Center: Expectations and Assumptions of ESL Students." In *Writing in Multicultural Settings,* edited by Carol Severino, Juan Guerra, and Johnella Butler, 220–33. New York: Modern Language Association.

Hemmeter, Thomas. 1994. "Live and On Stage: Writing Center Stories and Tutorial Authority." *Writing Center Journal* 15, no. 1: 35–50.

Hill, Heather N. 2016. "Tutoring for Transfer: The Benefits of Teaching Writing Center Tutors about Transfer Theory." *Writing Center Journal* 35, no. 3: 77–102.

Ianetta, Melissa. 2004. "If Aristotle Ran the Writing Center." *Writing Center Journal* 24, no. 2: 37–59.

Kennedy, George A. 2003. *Progymnasmata: Greek Textbooks of Prose Composition and Rhetoric.* Atlanta: Society of Biblical Literature.

Laycock, Joseph P. 2015. *Dangerous Games: What the Moral Panic over Role-Playing Games Says about Play, Religion, and Imagined Worlds.* Oakland: University of California Press.

LeCluyse, Christopher. 2009. "Medieval Literacy in the Writing Center." *Writing Lab Newsletter* 33, no. 10: 10–13.

LeCourt, Donna. 2006. "Performing Working-Class Identity in Composition: Toward a Pedagogy of Textual Practice." *College English* 69, no. 1: 30–51.

Lochman, Daniel T. 1986. "Play and Games: Implications for the Writing Center." *Writing Center Journal* 7, no. 1: 11–19.

Loveridge, Jordan. 2016. "'How Do You Want to Be Wise?' The Influence of the Progymnasmata on Ælfrīc's *Colloquy.*" *Advances in the History of Rhetoric* 19, no. 1: 71–94.

Mackay, Daniel. 2001. *The Fantasy Role-Playing Game: A New Performing Art.* Jefferson, NC: McFarland.

Mendelsohn, Susan Elizabeth. 2012. "Rhetorical Possibilities: Reimagining Multiliteracy Work in Writing Centers." PhD dissertation, University of Texas at Austin.

Missakian, Ilona, Carol Booth Olson, Rebecca W. Blank, and Tina Matuchniak. 2016. "Writing Center Efficacy at the Community College: How Students, Tutors, and

Instructors Concur and Diverge in Their Perceptions of Services." *Teaching English in the Two-Year College* 44, no. 1: 57–78.

Newkirk, Thomas. 1995. "The Writing Conference as Performance." *Research in the Teaching of English* 29, no. 2: 193–215.

Newkirk, Thomas. 1997. *The Performance of Self in Student Writing.* Portsmouth, NH: Heinemann/Boynton Cook.

Newkirk, Thomas. 2004. "The Dogma of Transformation." *College Composition and Communication* 56, no. 2: 251–71.

Nowacek, Rebecca S., and Bradley Hughes. 2015. "Threshold Concepts in the Writing Center: Scaffolding the Development of Tutor Expertise." In *Naming What We Know: Threshold Concepts of Writing Studies,* edited by Linda Adler-Kassner and Elizabeth Wardle, 171–86. Logan: Utah State University Press.

Roozen, Kevin. 2015. "Writing Is Linked to Identity." In *Naming What We Know: Threshold Concepts of Writing Studies,* edited by Linda Adler-Kassner and Elizabeth Wardle, 50–52. Logan: Utah State University Press.

Schaafsma, David. 1998. "Performing the Self: Constructing Written and Curricular Fictions." In *Foucault's Challenge: Discourse, Knowledge, and Power in Education,* edited by Thomas S. Popkewitz and Marie Brennan, 255–77. New York: Teachers College Press.

Scott, Tony. "Writing Enacts and Creates Identities and Ideologies." In *Naming What We Know: Threshold Concepts of Writing Studies,* edited by Linda Adler-Kassner and Elizabeth Wardle, 48–50. Logan: Utah State University Press.

Thompson, Isabelle, Alyson Whyte, David Shannon, Amanda Muse, Kristen Miller, Milla Chappell, and Abby Whigham. 2009. "Examining Our Lore: A Survey of Students' and Tutors' Satisfaction with Writing Center Conferences." *Writing Center Journal* 29, no. 1: 78–105.

Walker, Jeffrey. 2011. *The Genuine Teachers of This Art: Rhetorical Education in Antiquity.* Columbia: University of South Carolina Press.

Waskul, Dennis. 2006. "The Role-Playing Game and the Game of Role-Playing: The Ludic Self and Everyday Life." In *Gaming as Culture: Essays on Reality, Identity, and Experience in Fantasy Games,* edited by Patrick Williams, Sean Hendricks, and W. Keith Winkler, 19–38. Jefferson, NC: McFarland.

7

INSCRIBING THE MAGIC CIRCLE IN/ON/OF THE WRITING CENTER

Kevin J. Rutherford and Elizabeth Saur

In our recent experience opening a new university writing center, part of what motivated us was a sense that a writing center and its work exists in a unique space within the institution, with certain affordances and emphases on values and practices not necessarily in place in other areas of the university. Opening a new campus writing center can be an exciting and a frustrating process, one defined by the complications and successes that arise out of the interactions in both individual consultations within the center and the interplay between the center and other areas of the university. As Neil Baird and Christopher L. Morrow's discussion of the application of heuristics in writing centers in this volume demonstrates so effectively, writing center staff and administrators constantly find themselves playing different games at different times with different stakeholders, trying to introduce everyone to the rules and keep all those rules straight.

Drawing on game studies scholarship—and viewing the writing center as a space of play—this chapter argues that understanding these endeavors and experiences as "magic circles" (Huizinga 1971) can highlight the importance of recognizing the overlapping psychological, social, and material considerations of situating and administering a (new) writing center. Ultimately, we assert that making the rules of a writing center explicit—recognizing the boundaries we interact within and making all players and designers known to everyone involved in playing—creates a more equitable and just game. We argue that viewing the writing center as a magic circle serves to level the playing field for both staff and students; it creates a space separate from (albeit within) the university—a space that can operate by different, more equitable rules than those that create and enforce inequity in the systems outside the circle.

In particular, in our discussion, we emphasize that magic circles are dynamic, multifaceted, and negotiable; therefore, an underlying goal

https://doi.org/10.7330/9781646421947.c007

or ethos—clearly communicated to stakeholders—should be in place to set the stage for the negotiation of any individual policy or practice. In other words, we believe part of the responsibility of administrators is to be as forthcoming as possible about the rules that bound their magic circles, especially the philosophies that form a foundation for them, so all of those affected can understand how to navigate and whether to negotiate those rules.

DEFINING THE MAGIC CIRCLE

To situate our writing center experiences within the theoretical frame of the magic circle and in an effort to provide some context for our understanding of the term, we first outline the history and usage of the concept in prior scholarship. The magic circle is a long-standing concept in game studies. Although the precise nature and importance of the term is debated (as we discuss below), it is nevertheless a foundational frame for understanding the bounds of play—a way of theorizing where play begins and ends.

History and Trajectory of the Concept

In *Homo Ludens*, Dutch anthropologist Johan Huizinga (1971) describes key elements of play. For Huizinga, an activity is only playful if it corresponds to several shared characteristics:

- It is free (entered without coercion, but also "free" in the sense of "free time"—superfluous, in other words).
- It is not the "real world" but rather an escape from it (although it still has a "real" social function).
- It is secluded from day-to-day activity, bounded in both time and space; it creates order, and in fact, Huizinga goes so far as to say that it *is* order (9).

Taken together, these elements of play constitute the magic circle, which Huizinga (1971) describes as a separate, transitory, but fixed boundary between normal occurrences and the special state of people at play: "All play moves and has its being within a play-ground marked beforehand either materially or ideally, deliberately or as a matter of course. [Spaces within which play takes place] are temporary worlds within the ordinary world, dedicated to the performance of an act apart" (10). Huizinga sees these separate spaces as operating by different—and clearly demarcated—rules than does everyday life.

Huizinga initially uses the concept of the magic circle to delineate what is truly play. He spends much of *Homo Ludens* considering spoilsports, cheaters, professionals, and others he sees as bad-faith players. In other words, Huizinga's magic circle is a sort of litmus for determining whether something should count as play rather than a descriptor of all play experiences. As we discuss below, later scholars use the concept for different purposes than Huizinga seems to have initially intended. This sort of transition may be familiar to those in rhetoric and composition, where Lloyd Bitzer's (1968) idea of the rhetorical situation made a similar move from a framework to determine *whether* situations were rhetorical to a method for *describing* rhetorical aspects of situations. This caveat about the evolution of the magic circle as a concept is especially relevant for considerations such as coercion, lack of lasting consequence, and being apart from everyday life.

From its first moment in game studies scholarship, the magic circle concept began to diverge from Huizinga's purpose of delineating ideal play situations. The magic circle made its way into game studies in the early part of the twenty-first century, with Katie Salen and Eric Zimmerman (2003) relying heavily on it as a framework for understanding good game design strategies in their landmark *Rules of Play*. Salen and Zimmerman (2003) argue that "the magic circle of a game is where the game takes place. To play a game means entering into a magic circle, or perhaps creating one as a game begins" (95). Therefore, the magic circle is a special space for Salen and Zimmerman, and they see it as explicitly set apart from the real world, as did Huizinga. Unlike Huizinga, who makes no distinction between the boundaries of play versus the boundaries of games, Salen and Zimmerman describe the magic circle as something unique to rule-based games that may either be created beforehand and entered into or that may arise out of social interaction. Regardless, they see the concept as useful for defining the space within which game play occurs, which is fundamentally separate from normal activity and therefore "magical" as a result.

Game studies scholar and economist Edward Castronova (2005), in contrast, notes that although the concept of the magic circle is foundational to understanding games, Huizinga's (and Salen and Zimmerman's) emphasis on play as both distinct and separate from the "real world" and "an act apart" (Huizinga 1971, 10) is misleading. Considering separate spaces broadly, Castronova argues that the real and virtual meet with much less distinction between them in several areas (specifically, in markets, politics, and law). As he puts it, "The membrane between synthetic worlds and daily life is definitely there but also definitely porous, and this is by choice of the users. What we have is an almost-magic circle, which seems to have the objective of retaining all that is good about the fantasy atmosphere of the synthetic world, while giving users the maximum amount of freedom to manipulate their involvement" (2005, 159–60). In other words, for Castronova, the magic circle is a useful concept, but it can also belie the important connections between virtual spaces and real ones. While Castronova's boundary of play is socially negotiated, similar to Salen and Zimmerman's treatment, it does not fundamentally separate the players from their ordinary lives, instead allowing players to cross between virtual and real spaces (and to exist in both simultaneously).

Like Castronova, Jesper Juul's (2005) *Half-Real* resituates the magic circle, emphasizing the importance of recognizing virtual space's dependent relationship with the real world. As he puts it, "The fictional world of a game strongly depends on the real world in order to exist, and the fictional world cues the player into making assumptions about the real world in which the player plays a game" (168). Juul believes social interactions precede and create magic circles. We rely on our understanding of reality to inform our understanding of virtual spaces (and spaces otherwise apart from our normal lives); these bounded spaces always function within a preexisting context and are in some ways arising out of that context. In other words, Juul seems to see spaces of play as metaphors of interaction that depend on our understanding of the components of those metaphors, spaces that also require players to understand bounded spaces *as bounded* and to recognize those bounded spaces as emplaced and informed by their position within the real world.

Following Castronova and Juul, game studies scholar Mia Consalvo (2009) sees the concept of the magic circle as relying on "structuralist definitions or conceptualizations of games. It emphasizes form at the cost of function, without attention to the context of actual gameplay" (411). That is, Consalvo critiques Huizinga's concept of the magic circle because it not only emphasizes a separation from the real world but also rejects any connection that emplaces the game space in a particular social or political context. Huizinga's concept turns a game into an abstract series of rules rather than a dynamic and negotiated environment with real agents and stakeholders. Instead, Consalvo argues that social negotiation of a game space arises out of a particular social context and that social context can have real effects for both the formulation of the game and the approaches players take to it, above and beyond the rules that constitute the game itself. While Consalvo's formulation of the magic circle is a useful and nuanced one, it remains limited, especially in terms of psychological and material considerations—particularly for our purposes in this chapter.

The Magic Circle Expanded

Drawing on several of the above and other authors, Jaakko Stenros (2014) provides a significantly elaborated overview of the contentious history of the magic circle in game studies scholarship while simultaneously arguing for the relevance and usefulness of the magic circle as a concept. Rejecting the "strong boundary" hypothesis others see at work in Huizinga and Salen and Zimmerman,[1] Stenros (2014) divides the magic circle framework into three overlapping and dynamic boundaries of play:

- A state of mind that supports or enables play (the "psychological bubble");
- A negotiated, collaborative, social element of play (the "magic circle of play"); and
- The virtual or material space/time where play is intended to occur (the "arena") (sec. 8, para. 1).

In other words, rather than a singular magic circle, play consists of multiple magic circles that intersect and interact. For Stenros, each of these elements is worthy of analysis and cannot be easily collapsed into the others; an analytical approach to the magic circle should treat each in concert as well as individually.

In our analysis of the writing center as a magic circle that follows, we rely chiefly on Stenros's formulation of multiple overlapping magic circles, given that this form allows us to be more productive in our discussions of how the magic circle operates within the physical, mental, social, and institutional spaces of the writing center—that is, are we talking about the mind-set of consultants and students? The social space of consultations or the center itself in university life? The material or temporal boundaries of consultations and the center? Each of these things is important in and of itself, and Stenros's consideration of magic circles allows us to treat each in turn. We see the magic circle as a usefully complex concept that can help us consider how we can actively intervene in the physical, mental, and social conditions of the bound (if porous) space of the writing center.

With the utility of the concept in mind, we want to emphasize that because of the nature of interactions in the writing center—interactions that can involve coerced attendance, have lasting consequences, and are not actually escapes from other parts of students' lives—we deliberately rely on Stenros's ideas rather than Huizinga's. Effectively, we want to use this concept to highlight how the work of the writing center involves a negotiated sense of earnest play that arises out of particular material, social, and psychological contexts. We believe that working with Stenros's framework of the magic circle in mind allows us to move toward a level playing field—one that enables everyone to play the same game with the same rules, with those rules made clear, explicit, and transparent. In other words, we see the magic circle as a framework that can help establish a more just and equitable writing center.

CONSIDERING THE PLAYING FIELD

Prior to exploring the ways applying the lens of the magic circle to the space of a writing center can reveal tensions and areas for opportunity, we want to provide additional information to help establish an understanding of the context of our writing center. With the support of a generous start-up budget through the provost's office and integrated in a recently implemented college-wide five-year strategic planning steering committee, we developed and launched a new writing center at our college over the course of an academic year. We began with a small staff conducting writing consultations and making (nearly) all of the decisions: two graduate assistant consultants, two faculty consultants, and a writing center coordinator. In addition to training staff, developing a writing center philosophy, and establishing the role of the writing center on campus, we also worked with various programs and stakeholders on campus to build the material space for the writing center, implement the scheduling program and integrate other technology and equipment, and develop sustainable relationships with departments, faculty, and administrators.

Throughout our experiences navigating these specific institutional contexts, we first had to explain what a writing center is—the work we do, the existing disciplinary knowledge, the foundational philosophies guiding our practices—before either forging a connection or requesting services. While this is important work at any institution, our university context made clear to us early on that defining and articulating our boundaries would need to be our top priority; remarkably few individuals had heard of writing centers previously, and even fewer knew what they were. And so, whether making an argument to the purchasing department about getting approval to subscribe to scheduling software or making a case to the human resources department about using our already approved funding to compensate new interns, we had to outline the boundaries and define the rules of the writing center.

In other words, these negotiations served to reinforce how important it is to have a central vision of a writing center's philosophy in mind while communicating with stakeholders; however, this process of articulating our purpose to various audiences also simultaneously outlined the rules and boundaries of our magic circle. In short, we had to inscribe (on the institution) and describe (among our staff) the writing center's magic circle, with significant meta-commentary about the processes and values that make the center work and the kind of work the center does, in addition to communicating these boundaries and practices to various other stakeholders (students, faculty, administrators).

Negotiating the Boundaries, or "Rule Zero"

Given that we were starting our writing center from the ground up, there were many times when the decisions that needed to be made, shared, and understood were overwhelming—especially when trying to communicate these new policies and rules to new writing center consultants and various university stakeholders. With these complications in mind, in what follows, we explore how the tabletop gaming community's concept of "Rule Zero" can help writing centers navigate such challenges. We find this concept especially relevant in that such an understanding actively makes room for mutual negotiation among students and faculty and therefore establishes a more equitable space for play. Given that some students are coerced or required to visit writing centers and thereby do not fit the traditional concept of the space being a magic circle, our goal was to ensure that, regardless of what brought them to the space, all students would recognize their agency in and contribution to the playing field. We wanted students and consultants to recognize the ways they could collaboratively agree on the boundaries, purposes, and effects of work in the writing center through communication about those elements.

Rules and Meta-communication

In his discussion of the magic circle's social characteristics, Stenros elaborates on the degree of meta-communication surrounding play. In general, there is some sort of (implicit or explicit) communication that signals to all involved that an activity is play. A signal that the magic circle is being entered and is socially constructed and therefore negotiable is both common and an essential component of creating a sense of safety among players. As Stenros (2014) puts it, "Trust is a key element. Indeed, the idea that play and games are safe is deeply ingrained in the discourse of game studies and especially game design. It ties into the idea that play is separate from everyday life and actions taken during play bear few consequences beyond the play session . . . games are a safe platform to practise" (sec. 7, para. 4). Underlying this concept of the magic circle's social nature is a relationship between safety and the freedom to negotiate rules within the space of play. In principle, this shared responsibility to negotiate rules helps all players feel as though they have equal agency in how the game is played. While we may often think of rules as structures that constrict our ability to act, in this formulation they are the things that help us establish trust with other players and are put to use to emphasize and build that trust. Rules can be constricting, certainly, but only when they are adhered to without a recognition of the underlying context of trust that guides their implementation.

Rather than seeing rules as inherently restrictive, then, we should instead see certain articulations of rules as poorly developed or divorced from their larger context. As game studies scholar Ian Bogost (2007) argues, "Procedures found the logics that structure behavior in *all* cases" (7, original emphasis). We live our lives by following sets of moral, political, social, and physical rules; and so it is important that we develop and enact sets of rules that allow us to achieve our overarching purposes. For Stenros, a magic circle is in part a social contract, and there is a sort of preexisting rule (determined by context) that governs the creation of the rules that structure play. The underlying rule, by extension, must have some goal that later rules are intended to support. In tabletop role-playing games, this meta-rule concept is sometimes referred to as the "Rule of Fun" or "Rule Zero." Tabletop games frequently involve complex rule systems, but for many players the ultimate goal of these games is to engage in a kind of collaborative storytelling. In service to that goal, sometimes these narratively focused players agree to ignore the rules in favor of dramatic, interesting, or meaningful consequences for the story—with the understanding that doing so actually serves their personal underlying Rule Zero.

In other words, players can use Rule Zero as a way to deliberately work against preexisting rules that undermine their goals and expectations. In the case of our writing center, we recognized the potential of Rule Zero to help us construct our magic circle with an emphasis on supporting inclusion and creating welcoming sets of practices while also helping to work against the preexisting rules of the institution that existed outside the space of the writing center.

Writing Center Philosophy as Rule Zero

When we began to develop our Rule Zero for the writing center—the philosophy expressing the foundational objective around which all other rules of play would be constructed—we first scoured the internet for philosophies of other programs and institutions. This approach served multiple purposes, as it enabled us to have a sense of ongoing conversations and practices that were being conducted in the field while also helping to familiarize the staff with a wide range of possibilities of what a writing center can be. In other words, looking at the philosophies and rules that were enacted in the space of other writing centers helped us select how we wanted to construct the boundaries and rules of our own magic circle.

While the staff did not initially discuss the writing center in terms of games or magic circles during these early meetings, the underlying concepts were certainly present. Throughout these discussions, consultants consistently used words such as fun, playful, accessible, and safe to describe how they envisioned the writing center. They highlighted (literally and metaphorically) sections of other philosophies that focused on creating a welcoming space—separate from but within the institution—that would help students from all across campus work on anything to do with writing. As the conversations progressed, the parallels between their vision for the writing center and the possibility space offered by the concept of a magic circle became clearer. Recognizing the opportunity to offer up an additional theoretical framework for thinking about the writing center, Beth mentioned magic circles (almost in passing), and the idea took hold. Given the similarities between the consultants' ultimate goals for an inclusive, nonhierarchical writing center and the affordances provided by the concept of the magic circle, having such a metaphorical understanding seemed to help everything take shape. This was especially useful given that our staff consisted of the writing center coordinator, two faculty instructors who were ten to twenty years out of writing center work, and two graduate assistants who had never done writing center work (one of whom didn't actually have much awareness of what a writing center was). That is, developing our Rule Zero also helped the staff identify their own roles to play within the boundaries of our magic circle.

Working within a Possibility Space

To complement this training approach, in addition to looking at examples of other writing center philosophies, the staff was also given Stephen

M. North's (1984) foundational article "The Idea of a Writing Center" along with his later reflection, "Revisiting 'The Idea of a Writing Center' " (North 1994). While North's "Idea" is a mainstay of so much writing center scholarship that it has almost become parodic, it was especially relevant for our purposes given the context of our objective: essentially, we were seeking to define the rules of our magic circle, how we would construct our boundaries, and how we would be situated in relation to the rest of the university. We were in a rather rare position in that beyond our actual physical location in the university, we had full autonomy; North's (1984) emphasis on "what can happen" (433) in a writing center therefore helped frame our conversation in terms of capacity rather than solely in terms of necessity. In other words, we approached the creation of our Rule Zero by thinking about what our writing center *could* be in addition to what our writing center *needed* to be.

It was evident early in our discussions that all those involved in constructing the rules of the writing center were deeply invested in ensuring not only that the space would be made available and accessible to *all* students across campus but also that *all* students felt welcome in the writing center. Within these conversations, it was also clear that establishing this space as a just, equitable environment would help the staff feel both comfortable and capable in their new roles as writing consultants (a focus we explore later in this chapter). Similar to Stenros's understanding of the magic circle as a conceptualization that grants equal weight to the material, social, and psychological factors that influence the exchanges between players, we considered how we could account for the multiple facets of lived experience. That is to say, with this framework and ethos in mind, we considered the wide variety of beliefs, policies, and aesthetics we wanted to adopt, including both those that are standard across writing centers (e.g., session length, center philosophy, terminology) and those less frequently associated with the work of writing centers (e.g., what kind of plants we wanted to bring into the space).

At the heart of many of these discussions about who we were as a writing center and what we could be was also an inherent (sometimes implicitly, sometimes explicitly expressed) desire to work against the notion of the institution from within. That is, we wanted to exist as a space that resided in the university but that abided by a different set of expectations—a place where students could come, feel comfortable, and just hang out while talking about their writing with others who also cared about their writing without any fear of evaluation, judgment, or censure. To put it another way, we wanted to establish a magic circle where students wanted to come to play.

While there is ample scholarship exploring the work of writing centers as marginal (see, for example, the collection edited by William J. Macauley and Nicholas Mauriello [2007]), we necessarily and actively refused to identify ourselves as on the margins of the university. Instead, we adopted a stance similar to Elizabeth H. Boquet's (2002); that is, in responding to the perennial discussions of writing centers' locations on the margins of institutional structures and academic dialogue, Boquet argues that "rather than assuming that writing centers arise *from* the margins, exist *on* the margins, and are populated *by* the marginal, we might instead view writing center staff and students as *bastardizing* the work of the institution. That is, we might say that they are not a threat from *without* but are rather a threat from *within*" (32, original emphasis). In other words, we might recognize the magic circles of the writing center that exist *within* the larger context of the university and consider how the rules of play within their boundaries offer opportunity and capacity for the players in those spaces.

DEVELOPING PLAYERS' ROLES IN THE WRITING CENTER

As an essential part of defining the rules and boundaries for the magic circle of our writing center, establishing the roles the writing consultants would play was integral to ensuring that the space we created was both welcoming and equitable for everyone involved. As an example of the type of negotiations we went through in defining our roles, we can turn to the staff's decision to use the term *consultant* rather than tutor to describe their work in the writing center. In line with the many discussions about this choice of terminology (Bruffee 1990; McCall 1994; McQueeny 2001; Runciman 1990; Trimbur 1987), the staff unanimously preferred consultant more because we wanted to actively resist the connotations of the term *tutor* than because we actually liked the term *consultant.* As William McCall (1994) points out, "Whereas tutors are expected to know the correct answers and to proscribe the proper and rigid structures into which the students' thoughts must fit, consultants are perceived as supportive listeners who work flexibly with clients to help them achieve what they have identified as their goal" (167). In other words, in keeping consistent with our desire to level the playing field as we developed our magic circle—and to work directly against hierarchical associations—we opted for roles that framed student-staff interactions as collaborative and coequal.

Developing Roles for Graduate Consultants

As an important part of this nonhierarchical environment, it was critical that all members of the staff felt like equal participants with equal levels of competence. However, despite the numerous conversations they had about writing center pedagogy, the many articles they read, and the multiple affirmations they received, in the weeks prior to the writing center's opening, it was clear that the graduate assistant (GA) writing consultants were being asked to take on roles of which they were not entirely convinced. Being involved in the process of establishing the philosophy and policies of the writing center—the rules of play that defined our magic circle—helped the GAs become more familiar with what would be expected of them because of the very fact that they helped define those rules; however, unsurprisingly, they still weren't entirely comfortable with the process of writing consulting or with filling their roles as consultants. This was largely because they hadn't yet played their part. That is, while they had ample understanding of what the rules were and the theories and philosophies that influenced those rules, they hadn't yet participated in the game those rules governed.

In their exploration of how a consultant's various selves enter into and complicate the writing center space, Thomas "Buddy" Shay and Heather Shay (this volume) draw parallels between role-playing games and the work that happens in a writing center. One recommendation their theoretical understanding offers is for administrators to help make explicit connections between role playing and writing center practices to help tutors negotiate their roles more effectively. In a similar vein, in our own writing center training, we played around a bit to help consultants identify the roles they would be playing within the magic circle we had created; we conducted mock consultations, we acted out possible scenarios, and we talked through student writing as if working with the students who wrote it. In other words, we developed a sense of knowing the rules of the magic circle by playing through them a few times. By doing so, the GAs not only became more familiar with their roles as writing consultants, but this play also fostered a sense of safety that helped them adopt a playful mind-set moving forward.

As the GAs acclimated to their new roles, they also re-envisioned the role of the writing center in their professional lives. For example, in one of Beth's first get-to-know-you conversations with one of the GAs, the topics of availability and scheduling came up. With visible nervousness, the consultant hesitantly let Beth know that she was also working another part-time job but emphasized that it wouldn't interfere with her work in the writing center. The hours for the writing center hadn't yet been set, and given that it was already two weeks into the semester, Beth assured her that it wouldn't be a problem—they could work around it. As the schedule was set and hours were established, they did so without any problem. Then, about a week after opening the center to the campus mid-semester, after having gone through the initial training sessions and conducted her first consultations, the same GA walked into Beth's office with a smile on her face (and what might be described as a lighter sense of being). She excitedly announced that she had quit her other part-time job, that she was so excited about her role as a writing consultant that she knew what she wanted to do for the rest of her life, and that she could come in to work whenever they needed her. As it turns out, that consultant is now pursuing a PhD in composition and rhetoric with a research focus on student agency in second-language writing.

After having developed a foundation of understanding what it meant to play their roles as writing consultants, the GAs could then better identify how they fit into larger conversations of writing center practice and also how they stood in relation to the rules of the writing center's magic circle in contrast to the rules that governed the systems outside of it. That is to say, they could better assess how they might want to embrace or resist their roles in relation to the larger magic circle of the writing center and the institution as a whole. In discussing graduate students' experiences working in a writing center, Carrie Shively Leverenz (2001) posits that "part of the professional identity of most writing-center workers" is that "they are people who try to break down hierarchies between those who know and those who don't" (53–54). Before trying to break down these hierarchies (and if our GAs knew anything, they knew that was what they wanted to do), they first had to know that they were the ones who knew. In other words, they needed to identify, understand, and embrace their roles as writing consultants before they could work against them.

Developing Roles for Faculty Consultants

The process of identifying and playing their roles as writing consultants was a bit more challenging for the instructors. Whereas the GAs were starting the development of their consultant roles from a (nearly) blank canvas, the instructors had to first dismantle the authority inherent in their roles already granted by virtue of their positions in the university. The instructors' struggle with their new roles is understandable and relatable for many. As Muriel Harris (1988) advises to keep in mind when starting a new writing center, "Tutors are coaches and collaborators, not teachers" (sec. 2). In other words, not only did the instructors have to assume new roles in the writing center, they had to assume roles that were situated in direct opposition to other roles they played within the university. Time and time again, the faculty consultants found themselves in the coordinator's office explicitly expressing their frustrations with their tendencies to step into the instructor role they had been playing for years. They explained that since what they knew of the writing consultation session so closely resembled their experiences with one-on-one conferences with their own students, it was hard for them to differentiate one role from the other. This was especially difficult with the first handful of consultations; many times the instructor consultants would knock on the coordinator's door, asking "you got a second?" when they were feeling particularly uneasy about whether they were playing within the bounds of the magic circle as consultants or crossing the line into their roles as instructors.

Gradually, however, the instructors' difficulties with assuming their new roles as writing consultants within the circle of the writing center lessened. This happened in part due to their gaining more experience over time, similar to the way the GAs had to play out their parts in order to understand them. However, the instructors' transitioning between their roles also became a much more fluid process once they recognized the significance of our Rule Zero and their ability to negotiate the specifics of other rules as deemed necessary by context. In other words, once the instructors shifted their focus from prioritizing the individual rules of our writing center to relying instead on our philosophy to help guide their decisions, they were better able to navigate their roles as consultants.

As an example of such context-based negotiation, one of our writing center's faculty consultants succeeded in adopting a playful mind-set during a consultation about a month after the writing center opened. A student in psychology had scheduled a half-hour appointment to work on her senior thesis. When the session began, it captured Beth's attention because of how excited, energetic, and animated it was. It soon became clear that the consultant and the student were involved in some invested negotiation: they debated earnestly over content but also over word choice, paragraph breaks, point of view, and other issues. The conversation bounced back and forth as they both considered aloud how every choice related to and affected the piece overall and where they wanted to go next in the session.

One of the rules of our writing center is that students can schedule half-hour or hour-long appointments; however, after hitting the half-hour mark, the student asked if she could stay for another half-hour. The instructor consultant confirmed that she was available for that half-hour and then happily continued on with the session. As that next half-hour passed and the hour-mark approached, the consultant walked into Beth's office hesitantly requesting permission to continue for another half-hour. She expressed that she knew it was "against policy" but that they were "having such a great, productive conversation" that she didn't want to cut it short if at all possible. Beth reminded her of the writing center's philosophy (aka their Rule Zero) and asked the consultant if she was comfortable negotiating the rules with the student. The consultant smiled, nodded, and sat back down at the table to work with the student. They proceeded to finish up the last session block with fifteen minutes to spare and continued to laugh and talk about things unrelated to writing until the time was up. After the student had left, the consultant came into the coordinator's office, exclaimed "that was so much fun," and turned and walked out of the office. Through this and similar experiences, the consultant was able to recognize the purpose of Rule Zero as a guiding principle of the magic circle, as she ultimately felt comfortable playing around with the policy in favor of the philosophy.

While this serves as just one example of the play the magic circle of the writing center fosters, such moments can not only help staff become comfortable in their roles as consultants but can also enable them to help students. Ultimately, the faculty consultants were able to understand on a psychological, social, and material level how the magic circle of the writing center functioned as a space within (but also separate from) the institution. As such, while they were hesitant and concerned about their roles in the writing center when it first opened, by the beginning of their second semester, faculty consultants were laughing alongside students throughout sessions, offering students snacks from their bags, sharing personal anecdotes to help students relate to unfamiliar topics, and riffing and talking trash about favorite baseball teams. In other words, as the faculty consultants developed more familiarity with and confidence in the rules of the writing center—and as they came to recognize the safety the magic circle of the writing center offered—they were more able to adopt a playful mind-set within its boundaries. Once outside these boundaries, however, this safety no longer applies; the temporary nature of the magic circle can lead to conflict once players leave its bounds—an issue we explore more fully later.

THE MAGIC CIRCLES OF THE WRITING CENTER

As we have discussed, the framework of the magic circle consists not merely of a single barrier surrounding play (however porous). Rather, it is a series of overlapping and dynamic, negotiated boundaries that encompasses psychological, social, and material aspects. In the following section, we imagine how each of these boundaries was envisioned within the context of our writing center, as well as how these considerations might be put into use in writing center administration more broadly.

The Writing Center as Psychological Bubble

Stenros (2014) describes the psychological aspect of the magic circle as "personal, a phenomenological experience of safety in a playful (paratelic/autotelic) mindset" (sec. 8, para. 2). Within the context of the writing center, the consultants (GAs and instructors alike) needed to create and understand the boundaries of their experiences from a personal perspective, one that allowed them to adapt to and become comfortable in their new roles. In other words, they needed to draw a magic circle around the contexts of their own experience of earnest playfulness in the writing center environment to understand how they would interact with students within the space and how they would consider themselves as consultants. As Stenros (2014) puts it, the psychological aspect of the magic circle functions as "a 'border' around . . . experience that guides . . . interpretation of the situation" (sec. 8, para. 2). Fundamentally, while the consultants were adjusting to their roles, they began to approach the space with what they deemed the appropriate amount of playfulness, shifting their attitudes to accommodate their interactions within the physical and social space of the writing center.

This consideration is especially important given how transitory the work of a writing center is. That is, if not in the writing center, consultants might respond differently to the same piece of student writing than they would while on duty. As a pointed example, out of a scheduling necessity, a faculty consultant in our writing center found herself working with her own student. In conversation with Beth following the session, the consultant confessed that while she struggled to maintain her consultant identity and adopt the appropriate mind-set, the resultant clashing of her competing roles as teacher and consultant effectively inhibited the playfulness of the session. Interestingly, she explained that her main concern after the consultation was how the student would perceive her in the following class session. In other words, she was worried that the student would see her the next day as consultant rather than teacher and that boundaries would be blurred from then on. Similarly, Stenros (2014) recognizes that within a playful mind-set, "There is a private world, but it is not cut off from the real world. Like Goffman's interaction membrane, when properties from non-play world enter, they are transformed" (sec. 5, para. 4). Ultimately, this transformative psychological boundary allowed consultants to perform the work of the writing center *as consultants* and to see themselves as contributing to the play involved in that work.

We want to point out, however, that while the psychological bubble created by the magic circle of the writing center can be generative for both consultants and students, it can also give rise to potentially problematic conflict and resistance. In particular, one significant consequence of the magic circle's boundaries is that the rules established within that space no longer apply once a player leaves the space. Given that all magic circles exist within larger real-world contexts, conflict can arise when a player meets unexpected resistance and is no longer protected by previously established rules and safety.

One of our faculty writing consultants, for example, found herself "blindsided" (as she described it) by another instructor unaffiliated with the writing center. The instructor unexpectedly confronted the consultant in a hallway elsewhere on campus and began vehemently and aggressively expressing her distrust of the writing center as a new student resource on campus. The consultant was left speechless, as she was outside of the magic circle—no longer in a space of play—and unprepared to either discuss the issues the instructor raised or defend the work she believed in. She then went to Beth to process her (lack of) response and was visibly shaken, ultimately concluding that she had "let the writing center down" and that the other instructor "won" as a result. Although the specifics of the instructor's rebuke are the result of political and institutional influences outside the scope of this discussion, the effects her accusations had on the consultant serve to illustrate how the safety of the magic circle can potentially lead one to have a detrimental false sense of security upon leaving its boundaries. In other words, the rules and trust established within the magic circle no longer apply outside of that space of play, and it can be difficult to navigate resistance when one no longer feels safe.

The Writing Center as Space for Play

In addition to the psychological dimensions of the magic circle, Stenros (2014) details what he calls the "arena"—the "temporal, spatial or conceptual site that is culturally recognized as a rule-governed structure for ludic action" (sec. 8, para. 4). Like the psychological bubble, the arena helps players understand the bounds of the play space. In terms of the writing center, physical location and design help delineate the space, as does the idea of a block of time for appointments (although, as noted above, these, too, are negotiable). In our writing center, we focused on creating a space that functioned within an institutional setting but that also subverted institutional restrictions. In short, we attempted to create something that invited and facilitated play, a space that was welcoming, warm, cozy, and relaxed. We made the decisions we did to foster a particular variety of social interaction within the space and for the space itself to function as a kind of threshold into facilitating that feeling even outside the temporal and social bounds of consultations.

Along similar lines, Stenros (2014) asserts, "As the social negotiation of a magic circle becomes culturally established and the border physically represented, arenas emerge as residue of the playing . . . These sites are recognized as structures that foster play even when empty (and they can be constructed in ways that seek to foster playfulness), but they require use to be activated as the border of the magic circle remains social" (sec. 8, para. 4). That is, spaces develop their own character and associations but must still house particular kinds of social interactions in order to be used as intended. With these concerns in mind, we structured our writing center to sustain itself as a playful space, to build on its own developing social history, and to continually reinforce particular kinds of interaction through its design. For example, we worked with offices on campus to order specific kinds of furniture, couches, and chairs in which one could relax. Some of the first things we ordered were coffee pots and cups. We put tapestries on the walls to soften the institutional feel. The GAs were tasked with the decor decisions, as they undoubtedly had a better sense of "what the kids are into"; after a series of conversations with students who came to the writing center in the early days and the GAs' own decision-making processes, what resulted was a mix of string lights, succulents, and whimsy. In short, we designed the space in an effort to make students (and consultants) feel as relaxed and comfortable as possible, to create a clear separation physically from the more regimented space of the university (particularly the building housing the writing center, which also contains some elements of the university bureaucracy). We did this with the hope that the space would communicate some of the core philosophies arrived at by the staff and so that, in turn, those philosophies would be constituted by the space itself.

The Writing Center as Social Contract

While the space of the writing center exists as a magic circle within the institution, we would also like to consider the magic circle that is inscribed on the individual consultations themselves. An element of Stenros's (2014) magic circle framework, what he calls the "magic circle of play," is explicitly linked to social contexts and the degree to which those concepts are arrived at collaboratively by participants: "The magic circle of play is the social contract that is created through implicit or explicit social negotiation and metacommunication in the act of playing" (sec. 8, para. 3). For Stenros, the magic circle of play is predicated on a collaborative relationship between players to establish the rules under which interactions will happen. Importantly, this context is not a blank slate—participants bring their preexisting social contexts with them into this bounded space, and (as Juul [2005] and others above point out) the magic circle of play relies on its outside context to exist and be understandable.

In relation to individual consultations, students at our writing center often arrived without a clear understanding of the rules of engagement in this new context of play. While we tried to communicate our purpose and our rules to the campus, this was a slow-going process—especially as a new writing center. Therefore, some students had to rely on preexisting contexts (e.g., conferences with instructors) to frame how they approached the situation. Other students may not have had any referent they could draw on apart from their normal approach to social interaction. As per the social contract of the magic circle, our consultants then either implicitly or explicitly made modifications to the emerging social interaction to more effectively bound the space—to make sure they were both playing the same game, in other words. Within this framework, consultant strategies (and student responses) are necessarily driven by any explicit policies for consultation, but the collaborative and negotiated nature of any individual consultation will also rely on an interpretation of Rule Zero—the degree to which any particular rule can be put aside to accomplish larger goals. The magic circle of play in individual consultations, therefore, preexists the consultations themselves but is also consistently fluid and highly contextual, with consultants and students agreeing on the goals and practices employed.

Communicating the purpose and boundaries of the magic circle of the writing center to students is especially important—not only to facilitate students' experiences in the writing center but also to help establish and reinforce the writing center's purpose within the larger institutional community. If students are made aware of their part to play in an individual consultation, they will arrive to that social interaction with a greater sense of purpose and agency. When the consultants begin the social contract of interaction by asking "So, what did you want to work on today," a student who has been informed about the rules of the game will feel prepared and ready to respond and play their part. In addition, students will find the writing center a much less intimidating or threatening place if they have a better sense of what to expect from their experience; there is less uncertainty to worry about, fewer rules to learn, and more opportunity to engage in the playful part of the game. However, when students are not made aware of the rules or those rules aren't communicated clearly and effectively, not only will students be less likely to be willing to play, but they will also be less likely to show up.

While conveying the purpose and etiquette of a writing center is necessary at any university, this has been a vitally important—and particularly challenging—task at our institution, especially as the majority of the campus has little familiarity with what a writing center is. In part, our difficulty has been due to the boundaries of our circle evolving as we have constructed it. Rules we previously had not considered have suddenly become obvious, albeit not necessarily intuitive. For example, for the last two weeks of our first semester, we suddenly had to limit the number of hours individual students could sign up for consultations to one hour a day. Conversely, policies and procedures we thought would arise repeatedly have fallen out of relevance in our daily activities. We anticipated having to negotiate with far more students who were resistant to and adamant about fixing their grammar than we actually had to do. Further, as we continued to grow, our policies did as well. As an example, in our second semester of operations, our hours changed, driven by the needs of students and faculty on campus (including adding an entirely new day to the schedule), and we added three new staff members: one additional faculty consultant and two undergraduate writing center interns. We were excited about these developments and actively considered how they factored into and influenced our already existing boundaries in meaningful ways. However, we also wanted to ensure that we were consistently communicating our philosophy and rules to the rest of the campus community—even as they evolved—as part of our recognition that the magic circle of play surrounding the writing center (at the level of consultation and at the institutional level) should be fixed as well as porous.

CONCLUSIONS

The concept of the magic circle (especially as elucidated by Stenros) has helped us consider the complex ways the psychological, social, and material dimensions of writing center work overlap and collide and how these elements can be generative and supportive for the writing center when the rules of the magic circle are clearly expressed to, and the roles clearly understood by, the players involved. Whether communicated explicitly and literally through magic circle terminology and game play or addressed solely in theoretical and metaphorical ways, we believe the framework of the magic circle allows us to understand our writing center as a bounded space that both preexists and is animated by its participants and that is structured in a way that encourages productively playful interaction in service of helping students improve as writers.

We acknowledge that this conversation has been written from the perspective of administrators opening a new writing center, and thus it is written from a rather specific context. As a writing center becomes more established on a campus, these magic circles must necessarily also change. One way to think about how writing center administrators and staff can make these porous boundaries as generative as possible is to consider our prior discussion of Rule Zero as a starting point for both establishing and negotiating policies and procedures. As our experiences demonstrate, as our writing center progressed through its first semester on campus, consultants negotiated their own strategies for working through the rules of their interactions with student writers. Relying on their connection to the philosophy has allowed them to evolve as consultants, with the GAs becoming more comfortable negotiating (and resisting) their granted authority and the instructors becoming more comfortable relinquishing the authority they have grown accustomed to in other roles they play on campus.

Essentially, we firmly believe that a writing center philosophy is instrumental in laying the groundwork for the rules of play within the magic circles of the writing center—whether in generating training materials for new consultants, defining the approaches to the social contract of individual consultations, or managing the use of the space itself. Moreover, the writing center philosophy can help provide a much-needed context for a center's developing presence on campus and its growth as an organic process, with the philosophy concretizing the work of the writing center for faculty and administrators on campus. The philosophy, like all rules, must necessarily evolve and be negotiated; however, we believe it should be the foundation of any discussion about what the writing center *is* and *does* and how to communicate that work to other stakeholders.

As we discussed above, trust is thought of as integral for most types of play. Players must trust each other and also trust that the rules are fair and are arbitrated fairly. In other words, they must trust that they are participants in fair play (with all the connotations of that phrase intended). In the context of a writing center, this emphasis on trust is just as important. Students, staff, administrators, and outside stakeholders should have the rules—and their negotiability—communicated clearly, and a writing center philosophy that functions as a Rule Zero can aid in that communication. When resistance to notions of justice and equity inevitably arises, writing center staff and administrators can rely on clear and open communication of an underlying philosophy as a guiding principle. Part of what draws us to the magic circle as a concept is that the psychological and social boundaries are inherently negotiated and therefore capable of iterative improvement. In establishing a writing center with a philosophy focused on inclusion, transparency, and equity, we recognized that all other policies and practices were ultimately negotiable and that foregrounding those elements opened a space for conversation, even (or especially) if there might be an impediment to reaching those goals.

Of course, we also recognize that we began from the unique position of being able to develop our Rule Zero from the ground up, with a collaborative and supportive staff and a small network of allies in administrative positions on campus, which allowed us to (with relative ease) rely on recursive collaboration regarding our mission, goals, and philosophy. Additional research is necessary to apply the magic circle concept to more established (and potentially more fraught) campus environments, especially in terms of how the writing center as a magic circle interacts and overlaps with the magic circles of academic divisions, departmental allegiances, and institutional culture and history.

NOTE

1. Stenros believes that neither Huizinga nor Salen and Zimmerman actually argue for this kind of distinct separation between virtual and real spaces; he claims that they are very much aware of and advocate for the social effects of play and a porous boundary between play and normal activity.

REFERENCES

Bitzer, Lloyd. 1968. "The Rhetorical Situation." *Philosophy and Rhetoric* 1: 1–14.
Bogost, Ian. 2007. *Persuasive Games: The Expressive Power of Videogames.* Cambridge, MA: MIT Press.
Boquet, Elizabeth H. 2002. *Noise from the Writing Center.* Logan: Utah State University Press.
Bruffee, Kenneth. 1990. "Peer Tutors as Agents of Change." In *Proceedings of the National Conference on Peer Tutoring in Writing,* edited by Stacey Nestleroth, 1–6. University Park: Pennsylvania State University Press.
Castronova, Edward. 2005. *Synthetic Worlds: The Business and Culture of Online Games.* Chicago: University of Chicago Press.
Consalvo, Mia. 2009. "There Is No Magic Circle." *Games and Culture* 4, no. 4: 408–17.
Harris, Muriel. 1988. "Writing Center Concept." *International Writing Centers Association.* http://writingcenters.org/writing-center-concept-by-muriel-harris/.
Huizinga, Johan. 1971. *Homo Ludens: A Study of the Play-Element in Culture.* Boston: Beacon.
Juul, Jesper. 2005. *Half-Real: Video Games between Real Rules and Fictional Worlds.* Cambridge, MA: MIT Press.
Leverenz, Carrie Shively. 2001. "Graduate Students in the Writing Center: Confronting the Cult of (Non)Expertise." In *The Politics of Writing Centers,* edited by Jane Nelson and Kathy Evertz, 50–61. Portsmouth, NH: Heinemann.
Macauley, William J., and Nicholas Mauriello, eds. 2007. *Marginal Words, Marginal Work? Tutoring the Academy in the Work of Writing Centers.* Cresskill, NJ: Hampton.
McCall, William. 1994. "Writing Centers and the Idea of Consultancy." *Writing Center Journal* 14, no. 2: 163–71.
McQueeny, Pat. 2001. "What's in a Name?" In *The Politics of Writing Centers,* edited by Jane Nelson and Kathy Evertz, 15–22. Portsmouth, NH: Heinemann.
North, Stephen M. 1984. "The Idea of a Writing Center." *College English* 46, no. 5: 433–46.
North, Stephen M. 1994. "Revisiting 'The Idea of a Writing Center.'" *Writing Center Journal* 15, no. 1: 7–19.
Runciman, Lex. 1990. "Defining Ourselves: Do We Really Want to Use the Word 'Tutor?'" *Writing Center Journal* 11, no. 1: 27–34.
Salen, Katie, and Eric Zimmerman. 2003. *Rules of Play: Game Design Fundamentals.* Cambridge, MA: MIT Press.
Stenros, Jaakko. 2014. "In Defence of a Magic Circle: The Social, Mental, and Cultural Boundaries of Play." *Transactions of the Digital Games Research Association* 1, no. 2. https://doi.org/10.26503/todigra.v1i2.10.
Trimbur, John. 1987. "Peer Tutoring: A Contradiction in Terms?" *Writing Center Journal* 7, no. 2: 21–28.

8

RPGS, IDENTITY, AND WRITING CENTERS

Layering Realities in the Tutoring Center

Thomas "Buddy" Shay and Heather Shay

Five people, a mixed group of men and women, sit around the table. They only know a little bit about each other. Most of the conversation surrounds pop culture references, outlandish jokes, and a bit of slapstick comedy. They have gathered once a week for six months, and they are back again to pull off a caper. They lean in conspiratorially as they plan how they will raid the tomb, kill the dragon that has been harassing the nearby town, and abscond with all its ill-gotten gains. Obviously, they will not be doing it themselves. They instead are engaging in a game that is a mixture of improv acting and board game. Each of them plays a character. One, a woman in her early thirties who works as a high school guidance counselor, portrays a gruff male dwarf who wields a battle axe and is a devout priest to a god of law and order. Another, a man still in college, is preparing his character, an ancient elf who wields the power of the cosmos, to throw fireballs at any adversaries. The third, an older man who is a prison guard by day, plays a hobbit fond of picking pockets and disarming traps, while the fourth, a female college student and sorority member, plays a half-elf using the powers of charisma and music to control her foes. Finally, the fifth, a middle-aged man who writes computer code by day, controls the universe, setting the scene and portraying any minor characters or adversaries the group might encounter. They are preparing for their weekly role-playing game (RPG), and although the action of the game is imaginary, writing center professionals can still learn something from it for their own work.

Across town, a female college student, a senior majoring in English literature, sits across a table from another student, a male first-year who struggles with the requirements for his first composition course. He is nervous because he has gotten low marks on previous papers, and he

https://doi.org/10.7330/9781646421947.c008

has come for help with this paper, the midterm, in a last-ditch effort to raise his grade. The woman comes from a very different set of circumstances. English classes always came easy to her, and she very much enjoys helping others learn better ways to string ideas together into a coherent whole. The pair are engaged in a writing center consultation, and, whether they are aware of it or not, each of them is playing a role very similar to the roles played by the gamers across town. Each of them will also engage in a bit of improv as they discuss ideas in the paper and how to solve issues the student faces. The tutor will vary her approach throughout the session to make her lessons more effective, and the student will (it is hoped) work to put himself in the best frame of mind to learn from any critique he might receive.

Much research has been conducted to describe and evaluate the roles writing center tutors play while they work with students. Muriel Harris (1980) describes several prominent roles, such as coach, commentator, and counselor. Tutors, according to Harris, take on these roles to better facilitate student success across a range of abilities. Robert Brown (2010) demonstrates that tutors also take on hypothetical roles as generic readers, the professor, or another audience, such as a scholarship admissions committee. Andrea Lunsford (1991) and Thomas Hemmeter (1994) problematize this relationship by showing that the tutor role can, in fact, lend itself to unequal relationships with students. Terese Thonus (2001) demonstrates that these unequal relationships may, in fact, be just what professors are looking for when she notes that some faculty see the tutor as a surrogate professor. Further, work by Harry C. Denny (2010) shows how one's identity outside of work can commingle with one's work in the writing center. He argues, "Like identity movements, people in education and those in writing centers specifically must negotiate a common ground of self and Other, of audience and rhetorical purpose" (23). In other words, part of the work in writing centers involves tutors being flexible in the way they present themselves in order to meet their students halfway and better empathize with and educate them. As shown by Jessica Clements in her chapter in this volume, writing center professionals can help tutors better do this by invoking the experience of video game playing, as it gives them a chance to imagine themselves embracing different identities. Likewise, they can better do this by applying the identity work common to tabletop role-playing gamers.

While the identity of writing center tutor is complex and often requires tutors to alter their personas as they work with students, as shown in previous research, we argue that the situation is even more complex than previously mentioned. Not only do tutors shift when they work with students, they bracket their selves when they walk through the door to work. Essentially, writing center tutors take on three layers of identity while working: their personal selves, including all of their status characteristics such as their race, sex, sexual orientation, gender expression, social class, and the like; their professional selves as they perform the role of tutor among their supervisors and other tutors, which might encourage them to ignore or subdue the impact from their status characteristics; and their personas that they adopt and change as they work with different students. This same layering of identity has previously been studied among players of role-playing games using the sociological theory of dramaturgy, and lessons from that study will be applied to face-to-face tutors in the writing center. Specifically, as we suggest, a better understanding of these layers can make some difficult sessions and conversations go much more smoothly.

THE DRAMATURGICAL PERSPECTIVE

One common theoretical approach among sociologists in game studies is that of the dramaturgical perspective (Fine 1983; Goffman 1997; Shay 2017; Waskul and Lust 2004). Based on the work of Erving Goffman (1959), the dramaturgical perspective emphasizes that people continually show others who they are through social interaction. Later sociological research using the dramaturgical perspective examines identity work, which "is anything people do, individually or collectively, to give meaning to themselves or others" (Schwalbe and Mason-Schrock 1996, 115). While games are often depicted (in the broader culture) as inconsequential and distinct from more serious pursuits, sociologists studying identity work have shown that the way people act in games can be quite similar to how they act elsewhere, as demonstrated by the work of sociologists Dennis Waskul and Matt Lust (2004).

When thinking about how writing center tutoring is similar to games, one particularly useful analysis is that of Waskul and Lust's (2004) examination of the layers of identity among participants in tabletop role-playing games. Waskul (2006) describes a tabletop role-playing game as a recreational activity that "is not competitive, has no time limits, no score-keeping, and, aside from the death of the player's persona [character], has no infinite definitions of winning or losing" (20). Participants pretend to be fictional characters, called player-characters (or PCs), whose action they verbalize as they (usually) sit around a table, collectively imagining a fictional world (Fine 1983). The games involve at least two people who create PCs and one Game Master (GM), who develops the world the PCs inhabit. Using dice, participants determine whether intended actions succeed or fail in the imaginary world, generating a collective story as they play (Grouling Cover 2010).

In one instance, a good friend of ours, Tim, decides to join us in a tabletop RPG we've been playing for a few weeks: *Star Wars*, by Fantasy Flight Games. To begin play, he must first create the character (i.e., persona) he will portray in our tales of derring-do. Given the numerous facets that make up any real person, Tim will have to build his character from the ground up, making choices about who he will represent; some of these choices will be ones people cannot actually make in real life. For instance, he first selects his character's species, and he decides to make a Rodian (think Greedo from *Star Wars: A New Hope*, Episode IV), clearly leaving behind the fact that he is a white human being; however, he also makes a male character, reflecting his actual sex. Tim also chooses his character's career, which will determine roughly how the PC functions within the group. He decides he wants to play a stealthy character who is also good at fighting hand to hand. Thus he decides to make the PC a spy/infiltrator. Again, Tim makes a selection that does not represent him in real life, as he is a former elementary school principal with little expressed interest or ability to engage in physical violence. Now that the key decisions have been made, Tim customizes the character by, for example, making his character better at particular skills and choosing from a list of talents (i.e., special abilities) his character can perform.

Once the character is complete, Tim is able to join the game. At this point, Tim, who is already familiar with tabletop role-playing, will need to use both the player identity to demonstrate he can follow the rules of the game and his persona to portray his character effectively. In one scenario, the PCs have been tasked with liberating the victim of a kidnapping scheme who is being held in an old warehouse in a run-down part of the city. While others in the group want to have their characters go in with guns blazing, Tim talks them into letting his stealthy character scout ahead so they know what they are getting into. In doing so, he is both portraying his persona, as his character prefers sneaking around, and demonstrating his recognition (as a player) that combat can be dangerous and everyone's characters will survive better if they can plan out their attack, thus increasing everyone's involvement in the game. Since Tim's action has much dramatic potential, he also displays his player identity when he follows the game's mechanics and rolls his PC's Stealth skill combined with his Agility characteristic to see if his PC succeeds. Luckily, he passes the check. Thus Tim's PC makes it into the building unseen, and the GM begins describing what the PC sees as he skulks through the area. With information provided by Tim's PC (which as a player, Tim had to say his character relayed to the rest of the group), the frontal assault from the rest of the party is successful and the victim is rescued.

THE PERSON

Waskul and Lust (2004) distinguish among three layers of identities the participants hold: the person, the player, and the persona. The "person" is who the individual is outside of the game (2004). It refers to all the characteristics—demographic and role-related—of the person that are true of them in their everyday lives. For instance, many role-playing gamers are young adults, often specifically college students (Shay 2017). Other identities could be employee, son/daughter, friend, boyfriend/girlfriend, and so on. Generally, role-playing gamers are expected to bracket these identities by ignoring some details of who they are in certain social contexts when they play the game. Waskul and Lust (2004) show that their participants sometimes did not even acknowledge what identities their person consisted of because everyone had so successfully bracketed themselves.

The person of a writing center tutor likewise encompasses all the qualities they possess and all the roles they fulfill, mostly outside of work. It refers to the tutor's race, sex, sexual orientation, gender expression, social class, and religion, to begin with. However, the person also includes aspects of the tutor's self, such as participation in hobbies and sports, family structure, relationship status, and other work obligations. In short, the person of the writing center tutor includes anything that defines one as an individual.

Unfortunately, some identities (i.e., some aspects of the person) are considered valid or relevant to writing center work, while others are not. For instance, Neisha-Anne Green (2016) points out that her racial and ethnic identities are often seen as detriments to writing center work. Due to prejudice and discrimination, many white people who speak Standard American English often presume that people who do not or who do so with a different accent are uneducated, unintelligent, or both (Denny 2010; Green 2016). Writing center tutors such as Green, then, sometimes feel that to effectively do their job, they have to hide or downplay portions of their person that others, such as white tutors, would not have to ignore. In the end, most of our tutors have aspects of themselves that they feel are irrelevant or possibly detrimental to their day-to-day work in the center, and this requires tutors to bracket their identities while at work, just as Waskul and Lust (2004) describe players doing when they join gaming groups.

But doing so, as Waskul and Lust (2004) point out for role-playing gamers, is not always possible. External events intrude on role-playing games when players get phone calls during games; cannot make it to specific sessions because of work, school, or family obligations; or cannot separate out who they actually are from who their character is supposed to be. For example, Heather Shay, the second author of this chapter, is a feminist sociologist. When studying role-playing games for her dissertation, she could not alter the fact that she was easily categorized as the only female at the table (at least when the game began). Further, she could not make herself unconcerned with sexist rhetoric within the game. The same applies in the writing center. Middle-class tutors cannot suddenly make themselves similar to students recruited from disadvantaged areas (Wolff 2000), alter their age so they are not younger than nontraditional students they might work with (Craigue Briggs 2000), or erase their history of LGBT activism (Denny 2010) when they step into the center. Understanding these fixed identities and explicitly thinking about how their identity influences them as a tutor is critical. Similar to Clements in this volume, we argue that educating tutors about the ways their various identities impact their sessions can improve those very sessions.

THE PLAYER

The player, the second layer of identity among tabletop role-players, is the individual who knows how to play the game (Waskul and Lust 2004). This layer focuses on the part of the individual that is expected to know the rules for how to play role-playing games in general and for that specific game in particular. In addition, the player is expected to recognize when to apply game knowledge and when not to do so. For instance, Waskul and Lust (2004) discuss the phenomenon of meta-gaming whereby players use knowledge about how the game works that their personas do not actually have. When participants do so, others often think it diminishes their enjoyment of the game (2004). Players, then, display their familiarity with the game rules, so that layer is immediately relevant to the game.

While each tutor has a complex identity outside of work, inside the writing center they need to present only part of that identity in order to go about their daily tasks. Similar to what Waskul and Lust (2004) show role-playing gamers have to do, we argue that tutors need to present themselves as a player, someone who knows the rules of the game (in this case the policies and directives of the writing center) and someone who is able to apply those rules to a variety of workplace situations.

For example, each center has specific rules and policies that tutors must follow. Nancy M. Grimm (2009) describes the evolution of her writing center in terms that describe many of the places we work. For instance, it has long been held that tutors should not be too directive in their approach to student work. While there are debates about how directive a tutor should be and the best ways to teach students academic English, it is generally agreed that the student should maintain ownership of their work and that the tutor should not act as a coauthor. This rule puts tutors in the same spot as role players who are asked not to meta-game (Waskul and Lust 2004). Here, the tutors likely know, for instance, a better way to word a sentence. However, rather than tell the student what to put down (as a coauthor would do), they instead ask the student leading questions to help best correct the error. Thus knowledgeable tutors are often expected to separate their person (i.e., what they know how to do) from their player (i.e., what they should do for the student as a tutor) as part of their professional presentations of self while on the job. Doing so is part of being a good tutor, in much the same way that not using knowledge a player has (i.e., what their person knows) but a character (i.e., persona) does not is part of being a good role player. Elsewhere in this volume, Christopher LeCluyse elaborates on how being a good role player (and tutor) requires recognizing that this difference in knowledge can actually occur in two different directions—that of the player to the character and that of the character to the player.

While knowing the rules of the game is important for tutors, they must also know what pieces are in play. For instance, in the game, players need to be aware of which dice to use at which time, which figure represents their character, and which rule books to bring to the table. Likewise, tutors are asked to learn more about and remain aware of other resources on campus that can help students. Being a savvy tutor often involves referring students to these services. In the first author's center, we work closely with the library to ensure that students can get help with their research questions and the research process. We also have a bulletin board where campus clubs can advertise events for students. On occasion, we refer a student to a professor or a department head if there are issues the student faces inside the classroom. On rare occasions, we might refer a student to the campus counseling center as a result of a piece of writing. The tutors must know all of these other departments to ensure that students get the services they need in their quests to matriculate. Thus being a good tutor involves being aware of campus goings-on and where to go for more information, just as a good role player knows which book to use or which character sheet to bring to the table.

Beyond learning the game, however, peer tutors often undergo other changes in identity as a result of their writing center work. Denny (2010) describes the rhetorical space many students and peer tutors occupy as a result of working-class roots. Not raised to speak Standard English, these tutors must learn and model dominant expressions and ways of being to help students assimilate into the culture of the educated. In essence, tutors may feel the need to submerge their social class background in their pursuit of showing they are *real* tutors, similar to role-playing gamers who are asked to ignore their actual status characteristics as they play a game with people who may have very different status characteristics than themselves (Waskul and Lust 2004). Both Denny (2010) and Mandy Suhr-Sytsma and Shan-Estelle Brown (2011) describe the power dynamics inherent in the use of language in the center and how they can foster subtle oppression for students and tutors alike. While this tendency is problematic, peer tutors, like savvy role players, need to learn how to express themselves like players in the game. They often need to prove themselves as members of the discourse communities they hope to help students gain access to, just as role players often learn of optimum character builds or better ways to interact with the game environment through experience.

In fact, this awareness of discourse and the need to present as middle class within the academy is a topic of some discussion in writing center literature. Denny (2010) explains the need for tutors to become more fully engaged in the game and to teach students about rhetorical choices they can make regarding assimilation or resistance. Likewise, Grimm (2009) posits that as her writing center evolved from protecting Standard English to embracing linguistic diversity, the preferred tutors shifted from those rooted in the middle class and the academy to those who were able to successfully navigate multiple discourse communities. In effect, the shift went to an embrace of those who learned the game while at college and who were able to teach new students the rules and strategies required to play it well. The shift is akin to role players who have played a variety of different games versus those who only know and love a single RPG or those who have played a variety of character types (such as wizards, rangers, and fighters) versus those who always play the same character type. In the end, those with a broader experience base are often better players.

As can be seen, there are certain common traits, common identities, encouraged in writing tutors. Tutors are often not the same people outside of work that they are while in the center, even to the point of using different linguistic traditions with their peer groups than those they use with students on the job. To fill the role required for them to remain employed and to get along with co-workers and other students, tutors, like role players, are required to present themselves as knowledgeable players, capable of working within the social framework provided by the center and by its parent institution. Being aware of those shifts can help tutors make them, as shown later in the chapter.

THE PERSONA

But the most obviously relevant layer is the third—that of the persona. The persona is the character (i.e., the PC) the individual portrays in the fictional world. In the case of tabletop role-playing gamers, such personas are often fantasy characters such as dwarves or elves who can use magic or fight with swords. Within the context of the role-playing game, participants portray individuals different enough from themselves that they are clearly stepping outside of their own lives (Fine 1983; Waskul and Lust 2004). Although these portrayals are clearly fictional, participants are expected to effectively imagine and then verbalize that persona so it metaphorically comes to life for themselves and others during the course of the game (Shay 2013).

While the identity of the tutor as a professional is somewhat stable within writing center work, tutors often change roles as they work with a variety of students. Although the stakes of a writing center session are higher (for the tutors, professional competence and economic gain; for the student, passing a course or earning a good grade), this level of identity is akin to the persona played by RPG players when they portray different characters. For writing center tutors, the persona may shift a number of times during a day's work as tutors find the need to portray themselves differently with different students. Again, conscious awareness of these shifts can help tutors make savvy rhetorical choices in how they present themselves to their tutees.

As Harris (1980) pointed out, tutors often need to change hats. For instance, one student may be quite proficient with their writing but lack confidence because of past experiences. In this case, the tutor might portray himself or herself as a counselor, there to encourage the student to take risks with their writing and let them know that they are on the right track in spite of their misgivings. The student, in this case, may walk away from the session believing the tutor is a very upbeat, positive person in general, even though the same tutor may need to shift again with another student and describe a lesson to a student in the place of an instructor (Thonus 2001). In that scenario, the tutor may need to take a stronger stance and position himself or herself as an expert on writing. The student in this scenario, dealing with the same tutor, may describe the tutor as highly knowledgeable but somewhat emotionally distant. In each of these scenarios, the tutor did not change who they were as a person. Instead, they simply shifted the message as a rhetorical strategy to better help the students served. In essence, they changed their persona or portrayed a different character with the different students. They probably did so in part because of the stakes associated with failing to do so. Writing center professionals would not seem to be good at their job if they acted the same no matter what the student appeared to need.

Brown (2010) notes that tutors often portray multiple personas during sessions with students. As evidenced above, a tutor may take on the role of a professor, using information from the assignment sheet to guide the lesson; however, at other times in the session, the tutor may provide more general reader-response feedback to let the tutee know what a particular passage did for the tutor intellectually or emotionally. In this case, the tutor acts as both a specific surrogate audience and a more general reader, all while still acting as a competent player in the writing center by shifting strategies. This shift is similar to the way a player may have their character change direction mid-session based on the player's knowledge of the game and the larger narrative.

But just like role-playing gamers who get better as they play at getting into their character and who learn how not to merely imagine themselves in different game scenarios, tutors often need practice and training to improve their capacity to shift personas over the course of sessions. Hemmeter (1994), for instance, argues that "by approaching the peer relationship from a performance perspective—as something constructed, fluid, and involving power negotiations" (36), we can help tutors do just that. Hemmeter's (1994) work indirectly demonstrates the parallels between role-playing gaming and tutoring. He shows that having tutors tell stories about their tutoring experiences during training helps both the storytellers and those who hear them process their experiences as well as see how they and others successfully navigate the many different personas they often have to portray (1994). Role-playing gaming is, in essence, collective storytelling with elements of chance added in to make it a game (Grouling Cover 2010). As Goffman (1959) argues about games generally and Waskul and Lust (2004) argue about tabletop role-playing games specifically, the main difference between what real people do and what gamers do is the seriousness of the portrayal. That is also the benefit of using role-playing games rather than just storytelling as the framing device. Since telling a story about one's experience is premised on the teller providing the truth, people can mistakenly see it as essential that the story be serious and thus reflect their actual person. While that is not inherently problematic, it can make it harder for the teller to separate the layers of their identity (person, player, and persona) when doing so would be beneficial. By contrast, treating tutoring as similar to a role-playing game encourages tutors to see that they can tell stories about themselves as a player or a persona and not be disingenuous.

Tutors' ability to portray multiple personas can impact how well they do their job. It can also impact how well they feel they do (even when that may not match how well the tutee thinks they did). Gamers do the same when they think about how well they portrayed their characters (Waskul and Lust 2004) or when they judge other players as good or bad gamers based on how creative their character portrayal was (Shay 2013). Tutors, like role-playing gamers, may thus engage in meta-cognition about their personas when they choose what persona to portray at any given moment.

COMBINING THE LAYERS

Sociologists point out that an individual is always composed of all of their identities at once, even if all of those identities are not obvious or seemingly relevant to the situation (Collins 1990). The same is true of role-playing gamers, as Waskul and Lust (2004) show. While tabletop role players often work to keep the three layers separate, doing so is impossible and, in some cases, would end badly. Participants are expected to act as if their person is not relevant to the game because no matter who they are, they are portraying a persona who lives in another place (and often time). But participants recognize that they use who they are in the rest of their lives to fill in gaps about what their persona would do (2004). Similarly, without knowing the rules of the game (i.e., enacting the player layer), they cannot make their persona act successfully. A role-playing gamer who does not know, for example, that their character will die when they fall below a certain number of hit points is likely to get their character killed accidentally.

Likewise, within the writing center, tutors can struggle with how to bracket their selves at work and how to further bracket their selves in sessions. Often, difficult sessions with students arise in part because the tutor is playing the wrong character to mesh with that student or the tutor is relying too much on their person rather than staying within the guidelines of the center as a good player would. Consciously being aware of the differences in how one presents in the center allows tutors to make conscious, rhetorical choices about how they present, which improves center efficiency.

All of this is not to say that the three layers of identity should be completely separated and that one only needs to be mindful of one set of personal qualities at any given time. Instead, the argument is made that one should be mindful about what aspects of oneself one presents. Qualities of the person may indeed need to be shown to a student to facilitate academic and personal growth. Likewise, choosing a different persona in different sessions is not to create a fiction; rather, it is to make rhetorical choices about how to best create a bond with a given student so the tutor and tutee can have a good working relationship. Below are some examples of how applications of the theory can help writing center tutors and directors handle a slew of messy situations.

APPLICATIONS WITHIN THE CENTER

Understanding how tutors need to bracket and shift their identities to present themselves as both professionals and competent students (in the case of peer tutors) and how they further adapt their personas when working with individual students can be a valuable part of tutor training. For instance, introverted writing tutors may at first struggle with initiating conversations with students during sessions. However, when they learn that doing so involves taking on a role or presenting themselves in novel ways, they may find themselves more open to the experience. Also, thinking of the writing center as a sort of gaming space wherein all participants work together toward common goals can help them leave life behind them, as it were, and focus on the tasks at hand. As with the gamers, everyone needs to remain cognizant of what is going on at the table.

When we understand that identity is composed of multiple levels, we can help tutors with their work. One example occurred several years ago and involved Mary, a middle-aged part-time tutor with a master's degree in English who took the position as a stepping stone into full-time work after being a housewife and mother for much of her adult life. She was working with Brad, who often came into the center and was having particular difficulty with a paper he was writing for his developmental English class. He had been diagnosed with ADHD and was unable to organize his thoughts on the topic. Often, in this situation, the tutor would help the student list ideas and then discuss ways to connect them into a coherent whole, drawing lines and moving things around as they did so. However, something about Brad tugged at Mary's heartstrings, and so she ended up creating an outline for Brad to follow based on what she felt would work best for the assignment and effectively taking over Brad's paper. Brad ended up with little input into what he was about to write. She was responding as the person—who wanted to help, no matter what the expectations about appropriate help were—rather than as the player, who would, it was hoped, consider expectations about what constituted appropriate help given the situation. Furthermore, because the writing center occupied only one small room and Mary's next appointment was able to see what transpired in the session with Brad, she likewise tried to pressure Mary into telling her what was best rather than guiding her toward possible options she could choose between.

After the students left, the coordinator (Buddy Shay) spoke with Mary, who said she felt bad for Brad because he put so much effort into his work even though he still struggled. The pair discussed boundaries within the session and the need to continually follow best practices even if we wanted to do more. As a full-time equivalent (FTE) learning center, our guidelines from the state were very strict about being too directive during sessions. To our auditors, telling a student what to write about instead of guiding them to their own answers was the equivalent of giving a student the answer to an exam question. After all, we could not single out any students for special treatment.

Had we better understood the layers of identity, this boundary work could have proceeded more smoothly. Mary, forearmed with the theory, could have considered that while she as a person felt strongly that she should provide extra help, her self as a player needed to keep in mind the rules of the center and the restrictions on how the game is played. She would have been in a better position to mentally bracket herself. Moreover, she could have considered that in her role as a character within the session, she would be best served by showing compassion, working with Brad to overcome his frustrations, but ultimately letting him do much of the work during the session. Doing so would have prevented her next student from expecting her to do so much of the work. Such awareness would have helped her further bracket herself as a tutor and less as an editor or co-writer. Specifically, it could have served as a reminder to Mary that creating an outline for the student did not fit with the policies of the writing center that encouraged students to see that they, as writers, could make numerous choices for their paper and were not to rely on the tutor to tell them the supposedly right way to form the paper. Such rhetorical choices are similar to what role players experience during a game when they make choices for their character that may not be in league with what they themselves would do in that situation (for instance, someone who would normally run from trouble plays a fighter who would run toward it with an axe). Instead, she allowed her self as the person—a deeply caring person—to dictate her behaviors in the session without considering the other layers of self involved in writing center work. While merely knowing the three layers of identity could have aided Mary in this situation, framing those layers as akin to a role-playing game potentially lowers the stakes. Had Mary known she was playing a character like a gamer does, she would have been less likely to see her behavior as incongruent with how she saw herself. In other words, a tutor who sees themselves as deeply caring is less likely to think behavior that seems less caring is problematic if she does not see it as an actual reflection on who she really is. The situation, then, is similar to the way role-playing gamers often make choices in games that they themselves would not make in real life (for example, when a player chooses to have their character engage in violence they would refuse to commit in the real world) (Shay 2013).

Just as research on role-playing games and identity can help tutors maintain professional distance during sessions, it can help them better relate to students. At one point Buddy had a new professional tutor working for him, Blake, who was a recent graduate with a BA in English. He was working on developing his professional demeanor, and he struggled. Blake was not from an academic family or a middle-class family. His father owned a junkyard where Blake also worked part-time, and he had a sense of being an imposter among the other tutors, all of whom also had degrees and most of whom had grown up middle or upper-middle class. As a result, he overcompensated during his sessions, developing an authoritarian mien that some students felt was a bit stuffy. Some students went so far as to call him rude because he was so focused on presenting himself as an expert that he came off as unfeeling. Essentially, rather than thinking of how he presented himself to students as a persona, he was assuming that only a certain type of person could do well in this field.

The essence of the issue came down to Blake feeling that others would not accept that he was intelligent unless he portrayed himself as a know-it-all and an expert in the field. I knew Blake outside of the center from his attendance at a few local game conventions; he, like me, was a gamer (*Dungeons & Dragons*), and I knew he was very much into the *Game of Thrones* franchise. As we were working in Raleigh, North

Carolina, which is a technological hub, I explained that many of our students would find it easier to relate to a fellow geek than to an authoritarian tutor. Likewise, I let him know that revealing such personal details about himself would also make him more relatable to students who were not geeky because then they would be able to see that he was more than just a writing machine. In essence, I was helping him create a persona that he could use with students without being in any way deceptive and that would allow his sessions to run more smoothly. There was some pushback, but ultimately Blake decided to give it a try. After another week, Blake became more at ease in the center, and students left more positive comments after working with him. Essentially, Blake suffered from some role confusion because he felt he had to ignore himself as a person in order to present a savvy persona. I advised him to allow more of his person to become part of how he presented.

If we had had a better working knowledge of the theory, our conversation could have gone more smoothly. Just as in a role-playing game, whereby certain character types are a better fit than others (sometimes a game is designed to focus on social interaction; other times a game is designed to focus on combat), certain personas work better in some centers than in others. Without knowing it, my conversation with Blake may have helped him determine what general kind of character he wanted to play with students. Having knowledge of the layers of identity could have made Blake's transition more fluid, as he would have known upon starting the position that he would need to pick and choose between personal qualities and revelations to find a way to best relate to tutees. Thinking of those layers in terms of a game one plays while at work would make it easier to shift because it would make the stakes, which can be consequential for both tutor and tutee, seem lower. After all, if it's just a game, if one fails at one point, there will always be other opportunities for improvement later on. It may have also made doing so seem less like yet another daunting skill he needed to learn, possibly making him feel even more like an imposter among his middle-class peers, and more like something he would do for fun—pretending to be someone else in a role-playing game.

Part of being a good tutor and a good professional involves engaging in what sociologists refer to as emotion labor, and this labor can be more efficient if one understands the differences among person, player, and persona in the tutoring center. Emotion labor, according to Arlie Russell Hochschild (1983), is when people control or alter their actual emotions or their emotional displays to conform to the expectations of their work environment. Too often, tutor training manuals and advice ignore how central emotions are to working in a writing center (Lape 2008; Lawson 2015). But seeing writing center work as akin to a tabletop role-playing game can help tutors navigate the emotion labor necessary to successfully work with students even when their emotions are raging. For instance, there was an incident witnessed by the first author wherein Dr. Heinz (a pseudonym), a volunteer tutor from the English faculty, had a nasty exchange with a student. Dr. Heinz was known as a no-nonsense professor and a harsh grader, and she had difficulty shifting into the more nurturing posture used by most tutors. This posture generally works best because students often feel vulnerable asking for help (Mills 2011). Essentially, she was unaware that some personas that can work for professors—such as her seemingly emotionless instruction—may fall flat as a tutoring persona.

The session began innocuously enough. They worked together discussing the student's goals for the paper, and then Dr. Heinz started advising the student. However, the student did not accept this advice and thought a different technique might be better. Dr. Heinz reiterated her points and stated that the student would need to follow her advice. The student grumbled, and soon I had to intervene because the session was close to becoming a shouting match, demonstrating that Dr. Heinz was not succeeding at the emotion labor frequently required of tutors (Lape 2008; Mills 2011). I worked with the student instead.

After the session, I spoke with Dr. Heinz about the differences between teaching and tutoring and how students need to maintain ownership of their papers even if that means knowing that a student will not take a tutor's good advice. We discussed the need to work with the student, brainstorm with them, and gently guide them to progressing in their work. Had we had a better understanding of the layers of identity, the role changes involved in shifting from a teaching professional to a tutor would have been easier to swallow, and it would have been easier for Dr. Heinz to see that she needed to display other aspects of her identity here than she did elsewhere. Once comfortable with that shift, she would have been able to adopt a more open persona that would have made it easier for students to accept her advice. Moreover, it could have helped her manage her emotions, as she might have been able to see the student's unwillingness to take her advice as less of an insult to her and more of a different rhetorical choice made in the game of writing.

At the very least, if Dr. Heinz had reflected on the idea that there were multiple layers of her identity such as those present in gamers in a role-playing game, it could have helped her see her performance in that session differently. She could have considered that her role as professor makes her a player in one game, and her role as volunteer tutor makes her a player in a different game with slightly different rules. Reflecting on this might have made her more effective at the emotion management necessary to be a good tutor because it could have helped her recognize that she would not have been inauthentic when shifting personas. In the game as a professor, it would be expected that she have power over the student to make her voice heard in the student's paper. However, in the game as a tutor, the power structures are intended to be more level. Rather than thinking they are being fake or pretending to be someone they are not, tutors can recognize the demands the social context (i.e., following the rules and norms of the writing center) puts on them. Just as role-playing gamers recognize that they are not being inauthentic when they make two very different characters for different games that have different rules, tutors can recognize that they are merely playing by the rules of the metaphorical game when they shift personas in an attempt to help students deal with different writing challenges. In Dr. Heinz's case, she could have better understood that teaching and tutoring are different games with different sets of expectations.

In addition, by recognizing that they are also a person, they can (it is hoped) more effectively separate out their behavior as a tutor from who they are outside of work. Just like role-playing gamers who sometimes take the game more seriously than they anticipated or than others would like (Shay 2013; Waskul and Lust 2004), tutors may face difficult tutoring situations they struggle to let go of emotionally before the next tutee arrives or they leave for the day. If tutors are encouraged to remember that they are bracketing their person (i.e., identity) to some extent when they enter the setting, they may be more capable of doing the reverse—bracketing the persona they used with one tutee or their experiences as a tutor (i.e., player) when they move into other social situations that have different expectations of them. This could be especially useful for peer tutors who may interact with their tutees outside of tutoring as fellow students (whether in classes, as part of student organizations, or as friends).

By explicitly positioning tutoring as similar to a role-playing game, writing center administrators can help tutors not only deal with all three levels of their identity but also, as Waskul and Lust (2004) note, handle the situation when the levels break down. A tutor who is explicitly encouraged to see how they are shifting among the layers of person, player, and persona should be more equipped to recognize when their attempts to keep their person out of the session are failing and know what to do about it than will one who is not aware that their person is always present or who fails to recognize the distinction between how they present to any particular student and who they really are.

CONCLUSION

As shown here, looking to game scholarship can help us as writing center professionals as we navigate the social and rhetorical spaces we inhabit. Identity is complex, and beyond merely taking on roles in pursuit of excellence, we all go a step further and change our identities as we work with students and with each other. To act differently with different people, be they co-workers or tutees, is not necessarily to be fake or inauthentic. Instead, it can be an honest rhetorical strategy designed to help us build rapport with others and ensure that we send our messages in the best light possible.

As tutors play the game of work, they develop new insights into the rules of the college culture they inhabit and learn to take on an ever-growing variety of characters they may portray as they work with students. In the end, working in the writing center has much in common with gamers navigating fantastic worlds of imagination and adventure, and recognizing those parallels can enhance the experience of everyone in the writing center.

Helping tutors see those parallels, however, requires active effort on the part of writing center professionals. First, administrators and tutor trainers need to explain the three layers of identity to tutors. Second, they need to explain how those layers, originally developed to make sense of people playing tabletop role-playing games, apply to their work in the writing center. Since many tutors will be unfamiliar with role-playing games (and the dramaturgical perspective this layering of identity draws from), they will need practice applying them. This step is particularly important because tutors may think their work is much more serious than gaming and so think that what occurs in gaming is irrelevant.

Although there are likely numerous ways to help tutors see how approaching tutoring as akin to participating in a role-playing game will aid their work, one avenue we speculate would be particularly fruitful is that of self-reflection. To that end, here are some questions designed to facilitate tutors' shifts among person, player, and persona that could be integrated into reflective practice for tutors after a session has ended:

1. What parts of my private life did I want to leave at home today while at work? Why did I feel that way? What parts of my workday today do I want to leave at work? Why do I want to do so?

2. How was I able to demonstrate my knowledge of my position and duties while on the job? How am I a savvy player? What could I have done differently to show that I know the rules of the writing center game?
3. What aspects of my person were most valuable in my last session? Why were they so valuable? What aspects were least valuable? Why were they less valuable?
4. How did I effectively portray myself to my last student in ways that helped us connect and facilitated learning? How could I have done better?
5. If I work with this student again, would I adopt the same persona? Why or why not? How will I refine my persona for similar students?

Using these questions or others like them should help tutors see that they make rhetorical choices in the way they interact with others. By framing discussion about these choices as similar to pretending to be someone else in a fantasy game, writing center professionals can encourage tutors to try on numerous personas (i.e., characters) without thinking they are doing anything wrong. Writing center professionals can then, it is hoped, help tutors see that taking account of their person, player, and persona allows them to playfully experiment with potentially more effective ways to tutor the myriad students they are likely to encounter in a writing center.

REFERENCES

Brown, Robert. 2010. "Representing Audiences in Writing Center Consultation: A Discourse Analysis." *Writing Center Journal* 30, no. 2: 72–99.

Collins, Patricia Hill. 1990. *Black Feminist Thought: Knowledge, Consciousness, and the Politics of Empowerment.* Boston: Unwin Hyman.

Craigue Briggs, Lynn. 2000. "A Story from the Center about Intertextuality and Incoherence." In *Stories from the Center: Connecting Narrative and Theory in the Writing Center*, edited by Lynn Craigue Briggs and Meg Woolbright, 1–16. Urbana, IL: National Council of Teachers of English.

Denny, Harry C. 2010. *Facing the Center: Toward an Identity Politics of One-to-One Mentoring.* Logan: Utah State University Press.

Fine, Gary Alan. 1983. *Shared Fantasy: Role-Playing Games as Social Worlds.* Chicago: University of Chicago Press.

Goffman. Erving. 1959. *The Presentation of Self in Everyday Life.* Garden City, NJ: Doubleday.

Goffman, Erving. 1997. "Social Life as Game." In *The Goffman Reader*, edited by Charles Lemert and Ann Branaman, 126–46. Malden, MA: Blackwell.

Green, Neisha-Anne. 2016. "The Re-Education of Neisha-Anne S Green: A Close Look at the Damaging Effects of 'a Standard Approach,' the Benefits of Code-Meshing, and the Role Allies Play in This Work." *Praxis: A Writing Center Journal* 14, no. 1: 72–82.

Grimm, Nancy M. 2009. "New Conceptual Frameworks for Writing Center Work." *Writing Center Journal* 29, no. 2: 11–27.

Grouling Cover, Jennifer. 2010. *The Creation of Narrative in Tabletop Role-Playing Games.* Jefferson, NC: McFarland.

Harris, Muriel. 1980. "The Roles a Tutor Plays: Effective Tutoring Techniques." *English Journal* 69, no. 9: 62–65.
Hemmeter, Thomas. 1994. "Live and On Stage: Writing Center Stories and Tutorial Authority." *Writing Center Journal* 15, no. 1: 35–50.
Hochschild, Arlie Russell. 1983. *The Managed Heart: Commercialization of Human Feeling.* Berkeley: University of California Press.
Lape, Noreen. 2008. "Training Tutors in Emotional Intelligence: Toward a Pedagogy of Empathy." *Writing Lab Newsletter: A Journal of Writing Center Scholarship* 33, no. 2: 1–6.
Lawson, Daniel. 2015. "Metaphors and Ambivalence: Affective Dimensions in Writing Center Studies." *Writing Lab Newsletter: A Journal of Writing Center Scholarship* 40, no. 3–4: 20–27.
Lunsford, Andrea. 1991. "Collaboration, Control, and the Idea of a Writing Center." *Writing Center Journal* 12, no. 1: 3–10.
Mills, Gayla. 2011. "Preparing for Emotional Sessions." *Writing Lab Newsletter: A Journal of Writing Center Scholarship* 35, no. 5–6: 1–5.
Schwalbe, Michael, and Douglas Mason-Schrock. 1996. "Identity Work as Group Process." *Advances in Group Processes* 13: 113–47.
Shay, Heather. 2013. "I Am My Character: Role-Playing Games as Identity Work." PhD dissertation, North Carolina State University, Raleigh.
Shay, Heather. 2017. "Virtual Edgework: Negotiating Risk in Role-Playing Gaming." *Journal of Contemporary Ethnography* 46, no. 2: 203–29.
Suhr-Sytsma, Mandy, and Shan-Estelle Brown. 2011. "Theory in/to Practice: Addressing the Everyday Language of Oppression in the Writing Center." *Writing Center Journal* 31, no. 2: 13–49.
Thonus, Terese. 2001. "Triangulation in the Writing Center: Tutor, Tutee, and Instructor Perceptions of the Tutor's Role." *Writing Center Journal* 22, no. 1: 59–82.
Waskul, Dennis D. 2006. "The Role-Playing Game and the Game of Role-Playing: The Ludic Self and Everyday Life." In *Gaming as Culture: Essays on Reality, Identity, and Experience in Fantasy Games,* edited by J. Patrick Williams, Sean Q. Hendricks, and W. Keith Winkler, 19–38. Jefferson, NC: McFarland.
Waskul, Dennis D., and Matt Lust. 2004. "Role-Playing and Playing Roles: The Person, Player, and Persona in Fantasy Role-Playing." *Symbolic Interaction* 27: 333–56.
Wolff, Janice M. 2000. "Tutoring in the 'Contact Zone.' " In *Stories from the Center: Connecting Narrative and Theory in the Writing Center,* edited by Lynn Craigue Briggs and Meg Woolbright, 43–50. Urbana, IL: National Council of Teachers of English.

9

THE QUEST FOR INTERSECTIONAL AWARENESS

Educating Tutors through Gaming Ethnography

Jessica Clements

Writing center handbooks can create narrow categories for those who visit writing centers: "the writer with a learning disability," "the writer with a physical challenge," "the adult learner" (Ryan and Zimmerelli 2016, xiv). Handbooks do this not because the authors are unaware of writers' multifaceted identities but because educating tutors to meet the varied needs of any writer is a tall order. I argue that contemporary tutoring practice based on universal design and supplemental strategies meant to address a single slice of a writer's identity could be enriched by embracing an intersectional framework, a framework that honors the way forms of social classification are complexly interwoven and presented in a writer's identity and performed in their composing processes. While some writing center scholars consider the place of intersectional theory in writing center practice (see Ballingall 2013; Denny and Towle 2017; Geib 2017), it remains an underexplored area of growth. In this chapter, I investigate the affordances of intersectional tutor education as effectively accomplished through game studies methodology: a gamer's autoethnography.

I facilitated a gamer's autoethnography assignment in my one-credit ongoing tutor education course in fall 2018. By requiring tutors to build an avatar (that may or may not resemble themselves) and to participate in ethnographic analysis of their identity choices and cultural activities within the third space of a video game, I brought explicit attention to the inherent complexity of my students' identities as well as to the visible and invisible variables that comprise the identities and cultural performances of those they encounter, lessons tutors can transfer to daily engagement with new writers in the writing center.[1] Elizabeth Caravella and Veronica Garrison-Joyner in this volume provide a productively

https://doi.org/10.7330/9781646421947.c009

nuanced discussion of third spaces, claiming that "third space refers to a *position of power* where students do more than test the limits; they renegotiate them or at least image that they might do so . . . They can take a *calculated risk*, one they might never have considered otherwise" (emphasis added). I found that after engaging in a gamer's autoethnography, tutors (as evidenced in their reflective essays) are better situated to appreciate the intersectionality of all writers and will be (with appropriate ongoing encouragement and application) less likely to essentialize clients based on isolated traits.

My research aims to answer the question, What does it look like to practically engage with intersectionality in the education and supervision of writing center tutors? To do so, I first provide context for the field's understanding of intersectionality, discussing the term's emergence through the work of legal scholar Kimberlé Crenshaw (1991) and the contemporary attention it has received by writing center scholars such as Harry Denny and Beth Towle (2017) and Elizabeth A. Geib (2017). Next, to establish the place of video games in writing center discussions of diversity, equity, and inclusion, I draw on rhetoric and composition researcher Timothy Ballingall (2013) to demonstrate the connection between intersectionality and multimodal composition. I then describe and define ethnography as a research methodology inherently well suited to querying issues of culture, identity, and (multimodal) expressive choices within a discourse community, providing examples from rhetoric and composition as well as game studies. I explain my gamer's autoethnographic reflection essay assignment—highlighting its affordances and challenges for motivating intersectional awareness. I conclude the chapter by situating the gamer's autoethnography as one choice among many for broaching important discussions of intersectionality within writing center communities; I argue that the gamer's autoethnography is a particularly worthwhile choice given its potential to sponsor experiential, empathetic intersectional understanding that will lead to ongoing critical brave/r space engagement in writing centers' social justice work.

INTERSECTIONALITY AND WRITING CENTERS

If writing center practitioners are to be convinced to devote scarce resources to situating the exigence of and facilitating intersectionality education with tutors, then revisiting the term's origins is an important first step. Crenshaw (1991) is often credited with bringing the term *intersectionality* into mainstream academic conversation. In "Mapping the Margins: Intersectionality, Identity Politics, and Violence against

Women of Color," Crenshaw (1991) calls out contemporary feminist and anti-racist discourses as having failed women of color in particular given their intersectional identities and intra-group differences (1242–43). In that piece, Crenshaw situates "the need to account for multiple grounds of identity when considering how the social world is constructed" (1245) by drawing readers' attention to a case study of male violence against women and how women of color are disproportionately marginalized within both categories of gender and race. Crenshaw ultimately puts "intersectionality" in opposition to "anti-essentialism" as "a way of mediating the tension between assertions of multiple identities and the ongoing necessity of group politics" (1296). She advocates that intersectionality means not only arguing for multiple identities but also "negotiat[ing] the means by which these differences will find expression in constructing group politics" (1299). Arguing that we must find ways to give voice to the voiceless whose "identities are constructed through the intersection of multiple dimensions" (1299), Crenshaw implicates us all.

Crenshaw is not often cited in writing center scholarship, although field practitioners have long been concerned with issues of diversity. This has sometimes come in the form of discourse on learning styles (Johanek 1991–92), connected or not to learning disabilities (Neff 2011). This has often come in the form of discourse on working with non-native speakers of English (Harris and Silva 1993). This has more recently come in the form of explicit conversations about writing centers and diversity, such as *Praxis: A Writing Center Journal*'s 2007 volume 5, issue 1, special topic publication on "Diversity in the Writing Center" (as well as Okawa et al. 1991; Sloan 2003). Even more recently, writing center practitioners will find key resources for interrogating writing centers' ongoing scholarly conversations about diversity in *Facing the Center: Toward an Identity Politics of One-to-One Mentoring* (Denny 2010), *Writing Centers and the New Racism* (Greenfield and Rowan 2011), and *Performing Antiracist Pedagogy in Rhetoric, Writing, and Communication* (Condon and Young 2017); intersectionality, however, has been less thoroughly researched.

Despite this dearth of attention in writing center scholarship, Crenshaw's work has received significant attention in feminist game studies. Feminist game studies scholars have brought awareness to the reality that few women are involved in game design, resulting in the production of stereotypical female characters and game world contexts of little interest to women; interrogated the socio-cultural contexts and experiences of those female players and developers; and (heeding Crenshaw) highlighted the importance of embracing intersectional approaches to game studies, including issues of sexuality, ethnicity, race, class, and modern masculinity (Cassell and Jenkins 2000; Kafai, Richard, and Tynes 2016 quoted in Lukomski 2019, 6). Issues of class in particular provide a contemporary common ground for game studies and writing center intersectional interests.

A hallmark piece that explores the writing center field's relationship with intersectionality comes from Denny and Towle (2017). While the authors primarily advocate for making visible the struggles of working-class and first-generation students, they believe an intersectional approach to writing center practice, helping tutors "develop clearer eyes for the spaces they inhabit" and "the need to respect the inner and outer lives . . . of the students who come to the [writing center]," will benefit the field writ large (under "We Are the Change We Seek" [Obama 2008]). In other words, they focus on socioeconomic class as a particularly pressing aspect of identity that often remains invisible without due attention to intersectional performance. They warn of the lasting negative impact of tutors' un/spoken judgments levied at students rushing away from a tutorial to get to a second job or a tutor's impatience with a student who does not understand recurring comments from their instructor about colloquial language or heavy reliance on personal experience. Such microaggressions at the single tutoring session level can have a proudly damaging impact on an individual student and can contribute to problematic relationships between institutions and their working-class students.

Denny and Towle (2017) argue that writing center practitioners have the opportunity to influence such relationships in a positive way: by raising tutors' intersectional awareness, just as I sought to do with my autoethnographic reflection essay assignment. Explicit awareness of and cognizant attention to the seemingly small and previously perceived insignificant issues of intersectionality are critical to a holistically effective application of intersectional methodology in the writing center. Denny and Towle (2017) draw on the example of communicative microaggressions. Victoria McArthur (2019) explicates the "identity tourism" trap—"the problematic affordance offered by virtual spaces, allowing one to appropriate another culture or identity online"—of avatar building (39).[2] Such seemingly small transgressions too easily compound into irreparable damage for marginalized populations, which is why Denny and Towle (2017) contend that intersectionality must be "at the *forefront*—of our tutor training, of our data-collection and data-publishing, [and] of our programming" (under "We Are the Change We Seek" [Obama 2008], emphasis added). This includes, expressly, writing center scholarship.

Creating opportunities for dialogue in brave spaces—spaces that embrace challenge, risk, and discomfort with the goal of collaboratively producing genuine dialogue on diversity and social justice issues—will result in writing center scholarship that can "challenge or affirm what theories and practices work with a range of students, tutors, and writing center professionals" (Denny and Towle 2017, under "We Are the Change We Seek" [Obama 2008]). Writing center scholars need such dialogue to generate authentic stories from students, tutors, writing center practitioners, and other institutional stakeholders that will lead to hypotheses and research questions that will propel writing center scholarship forward into innovative practice (such as the autoethnographic reflection essay assignment) rather than leaning too heavily on tutor training guides that may potentially lock them into safe—unrealistically narrow—practices: "Our tutor training guides and theory monographs offer writing center practitioners little guidance on how we address critical dynamics pulsing in the background of sessions" (Denny and Towle 2017, under "Introduction").

Like Denny and Towle (2017), Geib (2017) takes issue with the sometimes overly dogmatic nature of contemporary tutor training guides, "prescribing solutions to assumed hypothetical problems" (vii). Tutor training guides are often highly structured, a set of rules to be followed in line with proven "best practices" (3); however, while these strategies "make sense on the page," they rarely account for tutors' dynamic interactions with real-world students "in their messy, complex identities that do not fit into categories and how-to guides" (4). That is, writing center practitioners often turn to handbooks as a productive *starting place* for structuring tutor education.[3] Importantly, it is not my goal in this chapter to condemn writing center handbooks; rather, as a common touchstone for the field, these handbooks, collectively, serve as an embodiment of my larger concern: the positioning of identity in these handbooks is limited, and, as a field, writing center scholars need to extend these introductory conversations and afford more dynamic opportunities for engaging intersectionality. Neil Baird and Christopher L. Morrow in this volume also note that "heuristics have a rich history in writing centers; however, they are primarily conceptualized as *static*" (emphasis added). On the contrary, "In game studies, a player's heuristics adapt as more information is learned about the game state. We argue that writing center heuristics can benefit from the dynamic nature of game studies heuristics" (Baird and Morrow, this volume). Game studies provides a dynamic, postmodern framework for incorporating intersectionality into tutor education.

Directly relating to the autoethnographic reflection assignment presented in this chapter, for example, SF. Luthfie Arguby Purnomo and colleagues (2019) propose "gamemunication," a methodology comprising "prosthetic communication ethnography of game avatars" (1). Purnomo and colleagues point out that traditional game studies ethnographies tend to rely on Hymes's SPEAKING approach (attention to settings and scenes, participants, ends, act sequences, keys, instrumentalities, norms, and genres), whereas a postmodern (and actor-network theory-specific) approach might complement and extend such traditional analysis by focusing on GAMING—gaming systems, attributes, mechanics, indexicalities, narratives, and geosocial systems (13). In other words, a postmodern approach that focuses on the complexity of contemporary communicative context by attending to the diversely intricate ways multiple nodes (including objects as actants) intersect to influence discourse would provide a more in-depth understanding of twenty-first-century performances of identity. For writing center practitioners, this means a more complete understanding of how tutor and client identity expressed in consulting discourse affects contemporary writers and their writing processes.

Indeed, Geib (2017) argues that handbooks that take a "postmodern approach," such as *The Oxford Guide for Writing Tutors*, come closest to approximating this intersectional engagement (7–8). A "postmodern approach" is one that provides a means for "hearing and/or viewing sessions where [we] might study the intersections of what makes writing center sessions successful or unsuccessful" (57). Even postmodern guides are limited, though, argues Geib, because "discussion on gender, race, social class, ethnicity, nationality, sexual orientation, religion, age, and disability is limited, if at all included" (59). A satisfyingly intersectional approach to tutor education would involve intersectional problem solving, accounting for the marked and unmarked identities that traverse the writing center (9). Ultimately, "There is not and cannot be a guide for moments like this because they are individual, complex, and include intersectional components of identity that cannot be reviewed beforehand" (54). Geib suggests that tutor training guides, rather, move toward embracing discomfort, asking new tutors to engage with "racism, discrimination, privilege, and prejudice" (67) in authentic ways.

Geib's (2017) call comes alongside a larger movement in the field to consider the affordances of—to use Brian Arao and Kristi Clemens's (2013) and Denny and Towle's (2017) terminology—writing centers as "brave spaces," spaces that are predicated on honest (at least uncomfortable, if not painful) engagement with issues of diversity and social justice. Consider Asao B. Inoue's Conference on College Composition and Communication (CCCC) March 2019 address in which he implored white audience members to grapple with their discomfort. The comments inspired emotionally charged, though critically productive, discourse on the writing program administrators' listserv (WPA-L), ultimately propelling forward the field's conversation about the intersection of race and place in new ways. The question remains, however: What does it look like to practically engage with intersectionality in the education and supervision of writing center tutors?

MULTILITERACIES, MULTIMODALITIES, AND INTERSECTIONALITY

One opportunity to more explicitly imbue intersectionality into tutor education comes from Timothy Ballingall (2013), who reminds us of the ways the multiliteracy and multimodality movements intertwine around issues of identity. He is quick to point out that it is not just a similarity but also an interdependence that exists between identity politics and multiliteracy: "The point of multiliteracy . . . is as much about access, difference, and rhetorical agency as it is about text forms" (1). Ballingall suggests that analogies can be used to explain the relationship to tutors. Consider a website, says Ballingall, a text that relies on multiple modes to communicate its message. It is comparable to an intersectional identity comprising multiple historically and socially constructed subject positions. Meaning in both instances is discursively negotiated through multimodal performance. Similarly, texts and identities may be recognized or excluded, marginalized or legitimated within given discourse communities and contexts.

Readers might consider, relevant to the tutor education assignment introduced in this chapter, the analogy of games or, more specifically, of avatar building. According to Mikael Jakobsson and colleagues (2019), academics who built a character design tool for game design industry leaders to visualize diversity and inspire creative reflection, "character design was chosen as the point of intervention both because of the crucial role it plays in relation to identification among players and because characters play a central role both in worldbuilding and the establishment of a repertoire of abilities that the player can access through the character" (Chess 2017; Shaw 2012; Taylor 2002; Yee 2006; all quoted in Jakobsson et al. 2019, 1). In each case, actors are making multimodal choices to communicate to a more or less informed audience: "The continual construction of identities depends on one's ability to communicate through 'multimodal performance' as well as on a discourse community's ability to recognize that performance as such with the aid of the community's 'cultural literacy'" (Ballingall 2013, 3). Ballingall suggests that writing centers should seek to be culturally literate centers, that writing center practitioners should be invested in bridging discussions of multiliteracy, multimodality, and intersectionality because it will turn the focus away from purchasing expensive hardware to fulfill ambitious multimodal goals and toward leveraging multimodal expression as a means of social justice (2).

Ballingall (2013) advocates that during a training course or staff meeting, writing center practitioners ask tutors to "think about the ways in which their cultural identities are communicated, or how they may interpret the communicated identities of others" (3). He cautions that it is the writing center practitioner's duty to effectively introduce the complicated concepts of multiliteracy and cultural identity, perhaps through analogy, but that asking tutors to share relevant personal experiences will help keep them engaged (3). He further suggests that reflection is necessary; writing center practitioners should give tutors at least a week to think about the ideas and consider them as they complete their tutoring work throughout the week—with the goal of contributing new examples to extended discourse on the same subject during the following class or staff meeting (4).[4]

Helping tutors better understand multimodality's social justice roots could lead to them using "multimodality and multiliteracy to rhetorically negotiate [their] subject positions both inside and outside of academic discourse" (Ballingall 2013, 5). Multimodality becomes less about crafting a slick-looking website and more about empowering writers with rhetorical agency by introducing them to and helping them interrogate the effectiveness of a multiplicity of composing tools and choices that might be used to reach intersectionally motivated goals. Such awareness-raising tutor education may open the door to the uncomfortably productive (brave) tutorials that Denny and Towle (2017), Geib (2017), and others call for. Enter the game studies–inspired autoethnographic reflection essay tutor education assignment.

ETHNOGRAPHY

Ballingall's (2013) theories about multimodal interrogation as a pathway to intersectional awareness in the writing center led me, a writing center and digital rhetorics scholar, to consider the possible affordances of gaming autoethnography as a means to raising tutors' awareness of intersectionality. More specifically, I piloted an autoethnography reflection essay assignment designed to get at the heart of Geib's (2017) question of intersectional empathy—"What would writing centers look like if both tutors and clients practice empathy more than they currently do (i.e., as a collective whole)?" (9) Given that ethnography is a research methodology that comprises a "deep dive into culture," one that utilizes participant observation to create an understanding of a group and "how" and "why" they do what they do (Collister 2012, under "What Is Ethnography"), my hypothesis was that tutors would develop not only awareness but also empathy after being guided in closely defining, describing, and reflecting on how identity performances manifest in the safe/r space of a virtual world.[5]

Ethnography is a methodology first (and perhaps primarily) embraced by anthropologists but one that has risen in popularity with game studies researchers. Ethnography is predicated on cultural immersion. Once excluded to external and physically located field sites, ethnographers now find value in researching offline as well as digital communities. Participant observation is one popular method that asks the researcher to function simultaneously as person participating and researcher analyzing (Collister 2012, under "What Is Ethnography"). The goal is to produce a thick description of the researcher's experience, relating facts, commentary, and interpretation of facts and commentary to readers to better help them understand the culture in which the researcher was immersed (Collister 2012, under "Finding Answers"). The gaming ethnographer has the advantage of logging all text chats and screenshots alongside their mental inventory of the experience. These texts, forum posts, audio recordings, images, and videos all may comprise data helpful in answering the questions of "how" and "why" (Collister 2012, under "What Is Ethnography"). More specifically, gaming ethnographers are interested in questions such as "Why do players do what they do," "How do players take the tools of a game and create a culture out of them," "How does their physical identity affect how they play," and "How does the game impact their lives and their identities" (Collister 2012, under "Questions for Games").

Gaming Ethnography

Furthering the scholarly conversation about ethnographic methodology's fit for game studies, Tom Boellstorff (2006) writes that borrowing from anthropology could "provide game studies with frameworks for theorizing culture" and could "provide a methodology—participant observation—for investigating games and culture" (30). Most interesting, Boellstorff describes participant observation in a way that approximates intersectionality: "The term 'participant observation' is intentionally oxymoronic; you cannot fully participate and fully observe at the same time, but it is in this paradox that anthropologists conduct their best work . . . In place of surveys or interviewing, participant observation implies a form of ethical yet critical engagement that blurs the line between researcher and researched" (32). Boellstorff describes it as a method "based on failure, on learning from mistakes to develop a theory for how a culture is lived—for its norms and its 'feel'—that may not be reducible to rules" (32). Similar to what Denny and Towle (2017), Geib (2017), and others suggest regarding the daily practice of writing center work, Boellstorff (2006) argues that participant observation is particularly apropos for game studies "where the object of study is emergent, incompletely understood, and thus unpredictable" (32). Even in these preliminary arguments regarding the utility of ethnography in game studies, Boellstorff suggests just how much researchers can learn from game cultures that might apply to real-world contexts. Boellstorff finds value in interrogating the relationship between "the metaverse and the physical world," asking questions regarding whether play is collaborative or individual, whether players choose avatars representative of their physical world identities, and the like. (33). In fact, he calls gaming a "master metaphor": "It appears likely that gaming and its associated notion of play may become a master metaphor for a range of human social relations, with the potential for new freedoms and new creativity as well as new oppressions and inequality" (33).

Many game studies scholars have since taken up Boellstorff's (2006) call to test the affordances of ethnographic methodology. In 2008, Kiri Miller brought attention to the rigorous input and valuable output of ethnographic methods in single-player worlds, suggesting that researchers may experience "shock of surprise and pleasure when an avatar does something the player at the controls did not quite have in mind" (under "A Tale from the Field"). Further, Seth Giddings (2009) reminded us that "the virtual space in this event of gameplay does not transcend the everyday and embodied; it is a real space to be explored and in which the player can act and be acted on. The virtual and the actual are both real, and in this event were each contained within the other, intertwining, each inflected by the other" (151).

In 2010, Bonnie A. Nardi wrote the now well-known ethnographic study *My Life as a Night Elf Priest: An Anthropological Account of* World of Warcraft. In chapter 2, she describes her methods, mentioning that "unlike research in most academic disciplines, where investigation proceeds according to scientific procedure involving hypothesis generation and testing, ethnography moves in a 'go with the flow' pattern that attempts to follow the interesting and the unexpected as they are encountered in the field" (27). Embracing dynamic interactions, she says, usually happens through one of two approaches: through application of theory (activity theory in Nardi's case) or "through the accretion of a multitude of details that impact a sense of the everyday texture of experience in a culture" (30)—a methodology much like what Denny and Towle (2017), Geib (2017), and others are calling for contemporary writing tutors to practice in support of intersectionality awareness.

Nardi (2010) takes up important topics such as entering a new culture and considering whether and how to make those you are observing aware of your status as a researcher. Even newer research, such as Diane Carr's (2019), reminds us of the shared underlying concern of (gaming) ethnographers, "an interest in centrality and marginality, normalized inequalities, privilege, and the 'othering' of particular social groups, all of which are considered significant because popular texts reflect the cultures they emerge from and because 'how we are seen determines in part how we are treated; how we treat others is based on how we see them' (Dyer, 1993, 1)" (708). Gaming ethnography as textual analysis, Carr (2019) tells us, "is a disorderly process that involves improvisation, iteration, and adaptation . . . Textual analysis can be *uncomfortable*" (717, emphasis added).

Rhetoric and composition scholars, too, have found value in avatar, identity, and ethnographic games research. A piece of particular relevance to this unfolding discussion of intersectionality and gaming ethnography is Samantha Blackmon and Daniel J. Terrell's (2007) "Racing toward Representation: An Understanding of Racial Representation in Video Games." In this piece, Blackmon interviews Terrell, a sixteen-year-old special needs student, "to ascertain whether Gee's learning principles can be extended to critical thought about the game environment itself and whether these principles can be applied to those who fall outside the spectrum of 'normal' cognitive abilities and practices" (204). Blackmon argues that "this identity development happens outside the game as well as[,] or instead of, inside the game environment" (208); that the player becomes the character *and* the character becomes the player. Blackmon references Michael Morgan to suggest that over time, fictional and "real" experiences become blurred in the player's mind and both become equally valuable resources from which to draw in making real-world decisions (209).

Blackmon and Terrell (2007) ultimately argue that we can and should use video games as tools for enhancing critical thinking (214). Video games offer "teachable moments" and interactions with underrepresented groups, a way to think about how women, minorities, and Others are rhetorically constructed (214). I argue that these groups may be considered intersectional others—and they may provide writing center practitioners with a way to think about how they might respond when presented with opportunities to discuss identity performance in the brave/r space of a contemporary multiliteracy center.[6]

An Intersectional Opportunity: The "Ethnographic Reflection Essay" Assignment

Drawing on gaming studies scholars such as Carr (2019), who suggests that "when the topic is the representation of identities or groups, it is not necessary to restrict the inquiry to certain parts of a game or to assume that different forms of meaning generation within the game should automatically be prioritized over others" (11), I presented sixteen undergraduate writing center tutors with an autoethnographic reflection essay assignment in late fall 2018 (see appendix 9.A). The assignment was introduced within an ongoing tutor education practicum in which we read Ballingall (2013) as a means for discussing intersectionality. Students were also provided with Denny and Towle (2017), which we discussed the following week—as well as early challenges students perceived with the assignment, including technical difficulties and time constraints. Chapter conclusions were drawn from grounded analysis of their final reflective essays.

Grounded theory is a systematic, inductive methodology that begins with a research question and the collection of qualitative data (i.e., in what ways, if any, did this ethnographic gaming experience enrich your understanding of intersectionality). The researcher reviews the data for repeated ideas or concepts (themes), which are tagged with codes. Researchers propose claims or theories based on patterns of codes (see Creswell and Creswell 2018). With an emphasis on both emergence and researcher autonomy within the overlapping frameworks of intersectionality and writing center research, I took a constructivist approach to this grounded analysis—noting that data, themes, codes, and similar factors are not merely discovered but are constructed by the researcher as a result of interaction with the participants (in this case, through the guided pedagogical exercise). In short, I carefully designed the autoethnographic reflective essay assignment with appropriate guiding questions based on an extensive literature review, then reviewed (and re-reviewed) students' autoethnographic reflective essays, noting and marking common words and phrases that spoke to collective themes and patterns regarding their expression of intersectional awareness pre- and post-ethnographic gaming experience. In this chapter, I make claims about what these patterns might mean for application in disparate tutor education contexts.

Student Reflections on Intersectionality

Fifteen of the sixteen writing center consultants completed the assignment by exploring a variety of gaming worlds: *World of Warcraft, Adventure Quest Worlds, Harry Potter: Hogwarts Mystery, Entropia Universe, Order and Chaos 2: Redemption, Overwatch, The Elder Scrolls Online,* and *Dungeons & Dragons.* While two of the fifteen students felt the exercise was less than beneficial (because, respectively, they did not experience enough interaction to draw conclusions or they felt it was irresponsible to draw conclusions about real-world behavior from fictional universes), all fifteen students confirmed the importance of intersectional education for not only writing center consultants but also global contemporary citizens. Through the grounded analysis of students' reflective essays, I found that consultants nearly universally agreed that the following awarenesses were heightened through the autoethnographic reflective essay exercise:

1. Players tend to create avatars or to choose characters that reflect their real-world identities, and they build virtual identities that will give them an advantage within the game world's context of play.
2. Presenting a masculine identity versus a feminine identity affords (not unsurprisingly) distinct advantages and often provokes expression of stereotypical assumptions from other players about one's skills, abilities, utility, and general belonging in a game world.
3. Players outwardly marked as novices or "newbies" are often left to fend for themselves within a given culture or community, which was described as a particularly frustrating experience.

To Be [Me] or Not to Be [Me]

The writing center tutors at my institution came to this assignment with a variety of interest levels and experience in both gaming and ethnographic research. Across the board, however, tutors reflected how they came to understand that an avatar's skin, clothing, and visible weaponry and from which "land" the avatar hailed often designated the player's potential worth to other players within a game world context, complicating their initial claims that they chose to create avatars that looked like them but they didn't know why: "I think of a friend of mine who spends real money on skins for a game called *DotA 2* (*Defense of the Ancients 2*). I could personally never imagine spending actual well-earned money on something [like that], but he realizes that certain skins present an aura of expertise and commitment to other players that one cannot achieve without buying said skins" (Tutor GM).

You Play Like a Girl

More specifically, the tutors had much to say about how others responded to their gendered skin in their game world. One tutor had some particularly revealing thoughts about the intersection of gender and age:

> As I suspected, however, Stevie as a character was not representative of the player behind her. During a flight with Stevie to a new part of the map, we got to talking about our lives outside of the game . . . I was not surprised that this female character was played by a man . . . but I was surprised at the age of the character. When I expressed my surprise . . . Stevie felt that he had to defend his identity as a gamer. "Gaming has no age," he said in a neutral but apologetic statement. I immediately clarified that I agreed but felt guilty that this man felt he had to defend his status as a gamer to me, a twenty-something male player whose identity as a gamer would never be questioned due to my demographic.
>
> In the writing center, I fall into a similar position. As a young, heterosexual, white male, I am a part of a majority on campus, and I have to be careful that my privileges do not intimidate, offend, devalue, or otherwise compromise the identity of my clients. Although the patrons of the writing center do not have avatars to hide their true identities behind, I also have to be careful not to make assumptions based on appearances and stereotypes. (Tutor J)

While Tutor J learned a memorable lesson about white male privilege connected to the intersection of gender and age as it was performed in *Entropia Universe*, Tutor SGH came to an equally revealing conclusion connected to the intersections of communal discourse conventions and gender in *Overwatch*:

> In a role I was unfamiliar with, I was nervous. I decided to ask what kind of strategy would be best for the game. I said, "Should we use dive comp?" Immediately, all conversation in the waiting area stopped, and a player's voice asked, "Are you a girl?" followed by another voice saying, "You should switch off, G." I have to admit that I was not shocked by this behavior. (Tutor SGH)

While Tutor SGH may not have been visibly marked as female, she violated discourse norms that caused multiple members of the game world to immediately label her as not useful and ask her to leave the group.

Tutor A further confirmed that outward markers of race ("skin") and gender within *World of Warcraft* afforded him distinct advantages as a male player performing a male avatar. Moreover, Tutor A's intersectional reflections on race and gender are of particular interest; he concluded his reflective essay by challenging the reader (me, as writing center director) to consider the role of the guiding reflective questions, alluding to bigger questions about the responsibility of text designers more broadly: "The game could merely be a reflection of our divided society as opposed to being a platform for reflection and education. For example, had I played this game without the prompt, would I have picked up on all of the elements I have presented in this essay, or would I be so desensitized to all the separation and inequality as to not even notice it?"

Tutoring the Newbie

Many of the tutors felt the frustration of being "new" to their game world communities, if not to the activity of gaming in and of itself:

> The trouble was, nobody was paying attention to me . . . I started asking people questions out of character. Eventually, a player in the general chat window started explaining some things to me [but] in the middle of all this, one person typed in the general chat, "Just look this s*** up before you start." Immediately two different people jumped on that player, telling them to not be a "gatekeeping p****." (Tutor T)

Some tutors explicitly connected their experience as novice gamers or being new to particular gaming communities to the experience of students visiting a writing center for the first time:

> I was coming into the game with a fear and lack of knowledge about MMORPGs [massively multiplayer online role-playing games]. I think clients often may find themselves in this same sort of mind-set when entering the [writing center]. They are afraid of the environment, feeling like they are not nearly as experienced or skillful at writing as they need to be . . . I honestly just wanted a guide to tell me what was going on and what my role was. (Tutor K)

Without necessarily directly applying lessons learned to a tutoring context, Tutor L suggested that their gaming experience mirrored the way "like connects with like" in real-world encounters: "People tend to connect with those they feel in common with rather than those who are different, and it is this interactive premise that the game is founded upon." Tutor P also remarked how their in-game experience confirmed real-world distrust of an individual asking too many questions, noting that others would say " 'You ask too many questions,' which was sometimes followed by 'Get away!' . . . Society doesn't always trust curiosity."

In a follow-up pedagogical opportunity, tutors could be prompted to further reflect on implications for tutoring practice. Fellow tutors might encourage one another to consider the marked diversity (or lack thereof) of the collective writing center staff. They might also further reflect on individual experiences, as did Tutor AK: "I started to question whether or not my appearance makes my clients feel comfortable and welcomed." Most important, such discussions could address how staff, as collective and individual actors, should consider the invisible and intersectional identities of themselves and their clients.

Bonus Level: In-Center Applications

Some of the tutors brought up starkly difficult but potentially productive questions about race: "I only encountered one gamer, and she did not do anything. There was no interaction or dialogue. Perhaps it was because my skin was tan and hers was more fair" (Tutor AC) and "If I was a white male, would I have an easier time of finding an avatar that looked like me" (Tutor A). Brave questions such as these require innovative problem solving, and tutors reflected about concrete steps for continued critical engagement with intersectionality in the writing center, including the following:

- Insisting that difficult conversations need to happen;
- Employing listening, asking questions, and observing with a heightened awareness of intersectional needs; and
- Learning from our past practices by reviewing client report forms or other documentation to raise individual awareness of how we may be un/intentionally pigeonholing clients.

Tutor GM and Tutor RP described this potential path to more productive conversation well. Tutor GM said, "Just as a character's stat-bar shows their levels of experience, revealing the tools they have available to them, a conversation with a client with simple questions reveals what tools they have available to them." Similarly, Tutor RP noted, "As the writing center strives to be more diverse and intersectional, I think the most important thing is to be observant. Just like in *OC2*, I was able to learn through watching and observing. Sometimes you need to take a step back and immerse yourself into understanding, then you can dive in and help a lot more effectively." Tutor SGH further iterated how the in-game autoethnographic experience caused her to promptly begin reviewing her client report forms for signs of intersectional disempowerment: "I unconsciously assumed that somebody at this level must be Korean, because Korean players are widely known as being some of the best

Overwatch players . . . However, when I made that assumption in-game, I realized that I must be making assumptions in other aspects of my work, even if they are brief thoughts." With complex and critical reflections such as these, I argue that writing center practitioners cannot afford to ignore this question: What do video games have to offer writing center practitioners committed to innovating their writing centers as inclusive brave/r spaces by forwarding intersectional tutor education curricula?

CONCLUSION: A CRITICAL PLAY GAME DESIGN FOR TUTOR EDUCATION

Critical play game design methodology privileges nonhierarchical participatory exchange in order to disrupt traditional game design processes and, more important, to disrupt the existing social realities offered by the majority of contemporary games, a term I purposefully invoke from Flanagan (2009, 256–57). In the spirit of critical play, I acknowledge that the most fulfilling tutorials are often ones in which the tutor learns as much as the tutee. I, too, learned something valuable about the evolution of my tutoring pedagogy practices through this autoethnographic reflective essay assignment. I learned that there is much at stake in a writing center practitioner's interrogation of the safe/r and brave/r space continua.

The Why: Embracing Safe/r versus Brave/r Space Programmatic Identity

As a reminder, Denny and Towle (2017) reference Arao and Clemens's (2013) definition of "safe spaces" as spaces that "work to make teaching and learning environments . . . more inclusive through consciousness raising with elective workshop participants, who learn to become more aware and sensitive to a variety of identities and issues" (under "From Safe Harbors to Brave[r] Spaces in Writing Centers"). According to Arao and Clemens (2013), the goal of safe spaces is "authentic engagement with regard to issues of identity, oppression, power, and privilege" (139) and involves training in using appropriate language and engaging in productive and respectful interactions. Safe space training is sometimes synonymously referred to as "sensitivity programming" (Arao and Clemens 2013, quoted in Denny and Towle 2017, under "From Safe Harbors to Brave[r] Spaces in Writing Centers").

Skeptics of safe spaces, including Arao and Clemens (2013) as well as Denny and Towle (2017), suggest that a truly safe space is an impossible standard, that spaces can only be more or less safe (Arao and Clemens 2013), and that genuine growth *requires* "challenge, risk, and indeed discomfort" (Denny and Towle 2017, under "From Safe Harbors to Brave[r] Spaces in Writing Centers"). In a brave space, then, the goal is not simply to talk sensitively about difficult issues—issues of intersectional diversity and social justice—but to understand them, to rise to the challenge of open and honest dialogue about them. In a brave space, ground rules are established, hot-button issues are discussed, and organic—possibly painful—conversation leads to productive learning. This conversation may be painful in that "brave space[s] require disclosure in order to work. To make a space 'brave' its participants must be willing to speak openly about their identities, biases, and experiences" (under "From Safe Harbors to Brave[r] Spaces in Writing Centers").[7] Denny and Towle argue that writing centers need to embrace programmatic identity as brave spaces to bring fuller attention to otherwise hidden aspects of intersectional identity (such as socioeconomic class status) that impinge on individual writing tutoring sessions. They also argue that the writing center is perfectly situated to make an institutional impact regarding diversity and social justice.

Although the writing center is perfectly situated to make an impact regarding intersectionality, writing center practitioners may not always be perfectly ready. In their autoethnographic reflective essays, tutors in my ongoing education course pointed out, on more than one occasion, that I had *not* emphasized intersectional empowerment or described the writing center as a brave space in their initial training but focused on creating a welcoming, safe space. I did this by emphasizing tutoring strategies that followed practical impulses tied to the time constraints of the average thirty- to sixty-minute consultation: "Our clients are not looking for a brave space. We do not have time in even an hour-long appointment to create one. So I concentrate on making a safe space" (Tutor RT). Writing center practitioners tend to be hyperaware of these temporal constraints and of the need to provide evidence of satisfied clients to administrative stakeholders. So, how do we efficiently pursue new paths that wed writing center theory and practice with game studies methodology when our time, energy, and other resources are already so taxed?

The How: Future Quests for Intersectional Awareness

The calls for raising intersectional awareness have been made (the why), but if writing center practitioners are going to effectively answer those calls, then we need practical and productive methods for doing so (the how). Application might begin by reading and discussing relevant theory (Ballingall 2013; Crenshaw 1991; Denny and Towle 2017) and implementing small action steps, such as displaying pronouns on center name tags. This action could be supplemented with inclusive teaching exercises, such as asking tutors at a staff meeting to engage in the University of Michigan's (2017a, 2017b) personal identity wheel or social identity wheel activities. But the task of re/framing writing centers as brave/r spaces invested in the work of intersectional social justice is arguably formidable. Asking tutors to be critical of oppressive practices and structures requires cultivating an environment of mutual respect and relationship building and attitudes of ongoing critical reflexivity—a commitment to making the invisible visible in sometimes uncomfortable ways. This challenges tutors to take on a double cognitive load: that of effectively grappling with multimodal texts and their writers in context as well as navigating complexly performed social identities and relationships. When we think of that social cognitive work as a literacy, however, we are reminded that game studies methodology has something valuable to offer us, a new and dynamically impactful way of addressing intersectionality in writing center work.

For decades, games have been seen as innovative sites for productively developing language literacies. Nasser Jabbari and Zohreh R. Eslami (2019) remind us that massively multiplayer online games (MMOGs) "provide socially supportive and emotionally safe (i.e., low-language-anxiety) environments that afford multiple opportunities for L2 [second-language] learning and socialization, which, in turn, help L2 learners to enrich their L2 vocabulary repertoire and enhance their communicative competence in the target language" (92). Similarly, I argue that MMOGs might offer tutors a lower-risk environment for tackling the exceedingly complex job of developing intersectional awareness as a social literacy that can boost tutors' confidence in practicing within writing centers as brave/r spaces. My case study assignment may serve as an innovative starting point for cultivating tutors' journeys into experiential, empathetic intersectional understandings that will lead to ongoing critical brave/r space engagement in writing centers' social justice work.

Game Over? Limitations

In many respects, the autoethnographic reflective essay assignment was a success. Several of the tutors produced verbiage that evidenced an intersectional party line, if you will. Tutor T stated that "a simple exercise like paying attention to the decisions of yourself and others in an MMO can really help to illustrate that people's identities don't exist in a vacuum but are instead part of an intersecting web of influences."

Similarly, Tutor P reflected that "I need to make sure that I hear each and every writer that comes to me, not just as who they appear to be and not just as a '[university] student' but as they show themselves to me." Tutor SJH said, "Showing a lack of awareness or care for individuals will only communicate that the writing center is majority driven and seeks the easiest possible solution." Yet many of the tutors focused primarily on what one might expect—language rather than intersectionality. Still other tutors stayed in the tried-and-true lane of anti-essentialism: "All of these aspects showed me a bit about how easy it is to stereotype someone based on an appearance" (Tutor AK) and "At some point well into the gaming process, I remember thinking, 'Is this how international students, or people from different economic backgrounds, or nontraditional students feel when they come to college' " (Tutor C).

The less-than-fully-satisfying outcomes of this pilot assignment could be attributed to cogent limitations. This was the first time the course had been proctored, let alone the assignment, and tutors were performing various levels of resistance to ongoing education as a whole. I also introduced the assignment late in the semester, when academic-related anxieties were already high, so those not predisposed to enjoy playing video games might have felt further resentment at being required to engage in an activity perceived to be frivolous. Timing serves as a limitation in another sense; ethnographic immersion generally takes a significant investment—far greater than two hours—to result in satisfyingly significant data. While lessons were learned, a longer investment over the course of the semester would likely have resulted in more clearly confirmed understandings and applications of intersectional approaches to writing center work. There were, as expected, some technical difficulties, and, finally, many of the tutors were new to ethnographic methods, the performance of which could have improved with practice and feedback over time.

Press Start to Continue

Addressing these limitations would likely make a replication, extension, or expansion of this assignment even more successful. And despite these limitations, I know that I can yet capitalize on perhaps the most consequential limitation of all: the need for follow-up discussion. Situating the exercise at the end of the semester meant I only had the opportunity to immediately follow up with tutors individually and in writing, which is not necessarily ideal for discoursing about pragmatic application of intersectional practice in writing center contexts. Perhaps this follow up involves reshaping the assignment to comprise weekly gaming auto/ethnographic engagement and reflective journaling, with careful (explicit) attention to the ethical responsibilities of a faculty member asking students to play games in online spaces.[8] Perhaps this involves using Ballingall's (2013) questions as a formal analytical framework for reflecting on a series of offline tutoring sessions (see appendix 9.B). Perhaps this involves tutors and directors collaboratively composing a tutor training guide grounded in intersectional theory that offers richly complex narratives about bravely intersectional writing center conversations. Perhaps it involves a scaffolded mixture of all three. Above all, it is about finding ways to pair traditional writing center concerns of rhetorical literacy with contemporary citizenship concerns of social literacy, to increase tutors' confidence in and motivation toward programmatic social justice and empowering twenty-first-century multimodal composers in their daily intersectional performances. What better place to start than the third space of video games?

Indeed, I have sound reason to believe the seeds of intersectional exigence have been planted:

> And what if that feeling of standing out and not knowing the rules wasn't based on something inconsequential like understanding an online game, but on where I came from or how I grew up or what my educational background had been? I'm not claiming to have complete understanding of that experience, because it isn't something you can understand fully unless you have been through it; however, I did get just a glimpse, *and I'd like to think that I have a little more empathy because of it.* (Tutor C, emphasis added)

If tutors feel empowered to empathize with their clients in more courageous ways (Geib 2017), then perhaps we move toward a hybrid safe/brave space:

> Now, no human is a blank slate, so I do not attempt to present that. My closest approximation is to revert to the white girl. Then, as I read their writing, as we talk, I can start to pick out where their identities show through and point them out. This is where I let bits and pieces of my identity drip through, merely as a contrasting element to their own. Again, my goal is to expose them to these differences without feeling confrontational. (Tutor RT)

In this hybrid space, our clients may feel welcomed to "state what difference [their] difference has made" (Crenshaw 1991, 1299), and we gain the disparate stories that Denny and Towle (2017) and others call for that will forward intersectional theory into profound intersectional practice.

NOTES

1. Mary Flanagan (2009) suggests that games represent a "thirdspace, a social space with its own social relations, struggles, and symbolic boundaries" (253).

2. One might argue that a gamer's autoethnography inherently falls prey to the limitation of identity tourism, a significant limitation that must be explicitly addressed in the facilitation of any such classroom assignment or workplace professionalization exercise.
3. Examples of popular tutor training guides include, but are not limited to, the following: *The Bedford Guide for Writing Tutors* (Ryan and Zimmerelli 2016), *The St. Martin's Sourcebook for Writing Tutors* (Murphy and Sherwood 2011), and *The Oxford Guide for Writing Tutors* (Fitzgerald and Ianetta 2015).
4. See appendix 9.B for Ballingall's reflective follow-up questions.
5. Anyone familiar with Julian Dibbell's (1998) "A Rape in Cyberspace" or the more recent #Gamergate controversy (see Massanari 2017; Mortensen 2018; Todd 2015) knows that virtual worlds and video game culture are far from proverbially "safe"; by "safe/r," I mean to invoke pedagogical theory that suggests virtual worlds can function as space where "learners can take risks . . . where real-world consequences are lowered" (Gee 2003, 67).
6. A fuller discussion of contemporary avatar, identity, and ethnographic games research by rhetoric and composition scholars is beyond the scope of this chapter. Relevant sources include, but are not limited to, the following: Colby 2013; McNely et al. 2013; Waggoner 2009; Warren 2013.
7. For historical consideration of safe and brave spaces, see Boostrom (1998); for relevant contemporary discussion, see Cook-Sather (2016); Green (2018).
8. That is, online interaction may expose students to aggression, profanity, sexism, and other negative elements. Although a detailed discussion of the ethical concerns of teaching composition theory and practice in online spaces is beyond the scope of this chapter, it is critical that we attend to ethical digital citizenship if we require students to participate in virtual worlds. Contemporary research suggests that only approximately one-fifth of composition teachers are "critical" users of technology (Mina 2019); therefore, we must necessarily remind ourselves of how we are ethically liable to prepare our students for such brave, though risky, engagement.

APPENDIX 9.A

ETHNOGRAPHIC REFLECTION ESSAY ASSIGNMENT

Ethnographic Reflective Essay

This (pass/fail) graded assignment contains two parts: (1) an ethnographic gaming experience and (2) an accompanying word-based reflection.

PART I: ETHNOGRAPHIC GAMING EXPERIENCE

You will play. More specifically, you will play at least two hours of an MMORPG (massively multiplayer online role-playing game: any story-driven online video game in which a player, taking on the persona of a character in a virtual or fantasy world, interacts with a large number of other players). *Lord of the Rings Online* (https://store.steampowered.com/app/212500/ The_Lord_of_the_Rings_Online/) and *World of*

Warcraft are two good options (https://worldofwarcraft.com/en-us/start), though any MMORPG will do. There is an advantage to the ethnographic researcher who is *not* already familiar with her subject, so please avoid games you have a vested interest in. That is, you need to pick a game and start from scratch.

As you are playing, you will take detailed notes about your playing experience. Detail some of your most interesting, revealing, and/or strange observations (cultural experiences) within the game. The thick description of ethnographic research is meant to immerse an outsider in the cultural world you experienced, reflective of behavioral patterns that might explain "why do players do what they do." You will use relevant description to answer the following reflective questions as deeply and clearly as possible. You won't necessarily know which observations and experiences are "relevant" until you've had a chance to process them, so the more notes you take while gaming, the better.

PART II: REFLECTION

In a reflective essay of approximately 4 (double-spaced) pages (probably not less, maybe more), answer the following questions:

1. What choices did you make in creating your avatar or selecting your character? Why did you make those choices (from what experiences, predilections, advice, etc., did you draw)? How might those choices, in and of themselves, communicate a "cultural identity"?
2. How did your performance of cultural identity in this particular "skin" impact your interactions with the gaming environment and the gamers you encountered within it (how were you received in the game culture and why do you think you were received that way)?
3. How did you interpret the cultural identities of others within the game world? What assumptions did you make about their place in the game culture and why did you make those assumptions? How were other gamers "positioned" in the game—by you, by other gamers, by the game environment, by themselves?
4. In what ways, if any, did this ethnographic gaming experience enrich your understanding of intersectionality, in general? More specifically, how might you apply what you learned about intersectionality through this ethnographic gaming experience to more effective tutoring practice in everyday writing consulting? Why does a heightened awareness and commitment to intersectionality matter in twenty-first-century writing center theory and practice?

While the reflection in and of itself does not have to be strictly argument-driven, it should be coherent (perhaps theme-based) and comprehensive in nature.

APPENDIX 9.B

BALLINGALL'S (2013) REFLECTIVE FOLLOW-UP QUESTIONS (MULTILITERACY, MULTIMODALITY, AND INTERSECTIONALITY TUTOR EDUCATION DISCUSSION)

1. "Which modes do consultants employ when discursively positioning themselves at the beginning of a conference, as a conference progresses, or when controversial topics emerge in a writer's paper?
2. "Under which circumstances might one mode serve better to position the consultant than another?
3. "Might there be alternate modes to more subtly or more clearly position oneself in relation to the writer?
4. "How might the consultant's discursive positioning change, based on a different set of available communicative modes, when the conference is in a synchronous or asynchronous online context, for example?
5. "How do writers utilize semiotic systems to continually transform intersectional identities both within their (multimodal) text and within the dialogue of the conference?" (5)

REFERENCES

Arao, Brian, and Kristi Clemens. 2013. "From Safe Spaces to Brave Spaces: A New Way to Frame Dialogue around Diversity and Social Justice." In *The Art of Effective Facilitation: Reflections from Social Justice Educators*, edited by Lisa M. Landerman, 135–50. Sterling, VA: Stylus.

Ballingall, Timothy. 2013. "A Hybrid Discussion of Multiliteracy and Identity Politics." *Praxis: A Writing Center Journal* 11, no. 1: 1–6.

Blackmon, Samantha, and Daniel J. Terrell. 2007. "Racing toward Representation: An Understanding of Racial Representation in Video Games." In *Gaming Lives in the Twenty-first Century*, edited by Cynthia L. Selfe and Gail E. Hawisher, 203–15. New York: Palgrave Macmillan.

Boellstorff, Tom. 2006. "A Ludicrous Discipline? Ethnography and Game Studies." *Games and Culture* 1, no. 1: 29–35.

Boostrom, Robert. 1998. "Safe Spaces: Reflections on an Educational Metaphor." *Journal of Curriculum Studies* 30, no. 4: 397–408.

Carr, Diane. 2019. "Methodology, Representation, and Games." *Games and Culture* 14, no. 7–8: 707–23.

Cassell, Justine, and Henry Jenkins, eds. 2000. *From Barbie® to Mortal Kombat: Gender and Computer Games.* Cambridge, MA: MIT Press.

Chess, Shira. 2017. *Ready Player Two: Women Gamers and Design Identity.* Minneapolis: University of Minnesota Press.

Colby, Rebekah Shultz. 2013. "Gender and Gaming in a First-Year Writing Class." In *Rhetoric/Composition/Play through Video Games*, edited by Richard Colby, Matthew S.S. Johnson, and Rebekah Shultz Colby, 123–36. New York: Palgrave Macmillan.

Collister, Lauren. 2012. "Ethnography and Gaming: A Short Primer." *Motivate Play.* http://www.motivateplay.com/2012/11/ethnography-and-gaming-a-short-primer/.

Condon, Frankie, and Vershawn Ashanti Young, eds. 2017. *Performing Antiracist Pedagogy in Rhetoric, Writing, and Communication.* Fort Collins, CO: WAC Clearinghouse.

Cook-Sather, Alison. 2016. "Creating Brave Spaces within and through Student-Faculty Pedagogical Partnerships." *Teaching and Learning Together in Higher Education* 18. https://repository.brynmawr.edu/tlthe/vol1/iss18/1/?utm_source=repository.brynmawr.edu%2Ftlthe%2Fvol1%2Fiss18%2F1&utm_medium=PDF&utm_campaign=PDFCoverPages.

Crenshaw, Kimberlé. 1991. "Mapping the Margins: Intersectionality, Identity Politics, and Violence against Women of Color." *Stanford Law Review* 43, no. 6: 1241–99.

Creswell, John W., and J. David Creswell. 2018. *Research Design: Qualitative, Quantitative, and Mixed Methods Approaches.* 5th ed. Los Angeles: Sage.

Denny, Harry, ed. 2010. *Facing the Center: Toward an Identity Politics of One-to-One Mentoring.* Logan: Utah State University Press.

Denny, Harry, and Beth Towle. 2017. "Braving the Waters of Class: Performance, Intersectionality, and the Policing of Working Class Identity in Everyday Writing Centers." *Peer Review* 1, no. 2. http://thepeerreview-iwca.org/issues/braver-spaces/braving-the-waters-of-class-performance-intersectionality-and-the-policing-of-working-class-identity-in-everyday-writing-centers/.

Dibbell, Julian. 1998. *My Tiny Life: Crime and Passion in a Virtual World.* New York: Henry Holt.

Dyer, Richard. 1993. *The Matter of Images: Essays on Representations.* London: Routledge.

Fitzgerald, Lauren, and Melissa Ianetta, eds. 2015. *The Oxford Guide for Writing Tutors: Practice and Research.* New York: Oxford University Press.

Flanagan, Mary. 2009. *Critical Play: Radical Game Design.* Cambridge, MA: MIT Press.

Gee, James Paul. 2003. *What Video Games Have to Teach Us about Learning and Literacy.* New York: Palgrave Macmillan.

Geib, Elizabeth A. 2017. "Writing Centers, Postmodernity, and Intersectionality: A Study of Tutor Training Guides." Master's thesis, Purdue University, West Lafayette, IN.

Giddings, Seth. 2009. "Events and Collusions: A Glossary for the Microethnography of Video Game Play." *Games and Culture* 4, no. 2: 144–57.

Green, Neisha-Anne. 2018. "Moving Beyond Alright: And the Emotional Toll of This, My Life Matters Too, in the Writing Center Work." *Writing Center Journal* 37, no. 1: 15–33.

Greenfield, Laura, and Karen Rowan, eds. 2011. *Writing Centers and the New Racism.* Logan: Utah State University Press.

Harris, Muriel, and Tony Silva. 1993. "Tutoring ESL Students: Issues and Options." *College Composition and Communication* 44, no. 4: 525–37.

Inoue, Asao B. 2019. "#4C19 Chair's Address." https://www.youtube.com/watch?v=brPGTewcDYY&feature=youtu.be.

Jabbari, Nasser, and Zohreh R. Eslami. 2019. "Second Language Learning in the Context of Massively Multiplayer Online Games: A Scoping Review." *ReCALL* 31, no. 1: 92–113.

Jakobsson, Mikael, Noah Houghton, Uche Okwo, and William Wu. 2019. "Visualizing Diversity: A Character Design Tool for Creative Reflection." Proceedings of Digital Games Research Association 2019. http://www.digra.org/wp-content/uploads/digital-library/DiGRA_2019_paper_378.pdf.

Johanek, Cindy. 1991–92. "Learning Styles: Issues, Questions, and the Roles of the Writing Center Tutors." *Writing Lab Newsletter* 16, no. 4–5: 10–14.

Lukomski, Jordyn. 2019. "Creating and Analyzing Values, Ethics, and Inclusive Design in Environmental Storytelling for Video Games." Master's thesis, Purdue University, West Lafayette, IN.

Massanari, Adrienne. 2017. "#Gamergate and the Fappening: How Reddit's Algorithm, Governance, and Culture Support Toxic Technocultures." *New Media and Society* 19, no. 3: 329–46.

McArthur, Victoria. 2019. "Making Ourselves Visible: Mobilizing Micro-Autoethnography in the Study of Self-Representation and Interface Affordances." *Journal of the Canadian Game Studies Association* 12, no. 19: 27–42.

McNely, Brian J., Paul Gestwicki, Bridget Gelms, and Ann Burke. 2013. "Spaces and Surfaces of Invention: A Visual Ethnography of Game Development." *enculturation: a journal of rhetoric, writing, and culture* 15. http://enculturation.net/visual-ethnography.

Miller, Kiri. 2008. "The Accidental Carjack: Ethnography, Gameworld Tourism, and Grand Theft Auto." *Game Studies: The International Journal of Computer Game Research* 8, no. 1. http://gamestudies.org/0801/articles/miller.

Mina, Lilian W. 2019. "Analyzing and Theorizing Writing Teachers' Approaches to Using New Media Technologies." *Computers and Composition* 52: 1–19.

Mortensen, Torill Elvira. 2018. "Anger, Fear, and Games: The Long Event of #Gamergate." *Games and Culture* 13, no. 8: 787–806.

Murphy, Christina, and Steve Sherwood, eds. 2011. *The St. Martin's Sourcebook for Writing Tutors.* 4th ed. Boston: Bedford/St. Martin's.

Nardi, Bonnie. 2010. *My Life as a Night Elf Priest: An Anthropological Account of* World of Warcraft. Ann Arbor: University of Michigan Press.

Neff, Julie. 2011. "Learning Disabilities and the Writing Center." In *The St. Martin's Sourcebook for Writing Tutors,* edited by Christina Murphy and Steve Sherwood, 249–62. 4th ed. Boston: Bedford/St. Martin's.

Obama, Barack. 2008. "Barack Obama's Feb. 5 Speech." *New York Times,* February 5. http://www.nytimes.com/2008/02/05/us/politics/05text-obama.html.

Okawa, Gail Y., Thomas Fox, Lucy J.Y. Chang, Shana R. Windsor, Frank Bella Chavez Jr., and LaGuan Hayes. 1991. "Multi-cultural Voices: Peer Tutoring and Critical Reflection in the Writing Center." *Writing Center Journal* 12, no. 1: 11–33.

Purnomo, SF. Luthfie Arguby, St. Lukfianka Sanjaya Purnama, Lilik Untari, Nur Asiyah, and Novianni Anggraini. 2019. "Gamemunication: Prosthetic Communication Ethnography of Game Avatars." *Malaysian Journal of Communication* 35, no. 4: 1–16.

Ryan, Leigh, and Lisa Zimmerelli. 2016. *The Bedford Guide for Writing Tutors.* 6th ed. Boston: Bedford/St. Martin's.

Shaw, Adrienne. 2012. "Do You Identify as a Gamer? Gender, Race, Sexuality, and Gamer Identity." *New Media and Society* 14, no. 1: 28–44.

Sloan, David. 2003. "Centering Difference: Student Agency and the Limits of 'Comfortable' Collaboration." *Dialogue: A Journal for Writing Specialists* 8, no. 2: 63–74.

Taylor, T. L. 2002. "Living Digitally: Embodiment in Virtual Worlds." In *The Social Life of Avatars: Presence and Interaction in Shared Virtual Environments,* edited by R. Schroeder, 40–62. London: Springer-Verlag.

Todd, Cherie. 2015. "Commentary: Gamergate and Resistance to the Diversification of Gaming Culture." *Women's Studies Journal* 29, no. 1: 64–67.

University of Michigan. 2017a. "Personal Identity Wheel." https://sites.lsa.umich.edu/inclusive-teaching/2017/08/16/personal-identity-wheel/.

University of Michigan. 2017b. "Social Identity Wheel." https://sites.lsa.umich.edu/inclusive-teaching/2017/08/16/social-identity-wheel/.

Waggoner, Zach. 2009. *My Avatar, My Self: Identity in Video Role-Playing Games.* Jefferson, NC: McFarland.

Warren, Katherine. 2013. "Who Are You Here? The Avatar and the Other in Video Game Avatars." In *Rhetoric/Composition/Play through Video Games,* edited by Richard Colby, Matthew S.S. Johnson, and Rebekah Shultz Colby, 33–43. New York: Palgrave Macmillan.

Yee, Nick. 2006. "Avatar and Identity in MMORPGs." *Daedalus Project.* http://www.nickyee.com/daedalus/gateway_identity.html.

10

I TURNED MY TUTOR CLASS INTO AN RPG

A Pilot Study

Jamie Henthorn

Writing center directors have obligations unique to their individual institutions, but one task they tend to share is educating tutors. While much of tutor education focuses on practical day-to-day issues, directors also invite tutors into the narratives that surround writing center work. Tutor courses vary significantly in duration and structure, making it challenging to create a homogenized understanding of what tutor education should be. While there have been many studies on tutor education—Karen Gabrielle Johnson and Ted Roggenbuck's (2019) collection in *WLN Journal* is one of the more recent—this collection highlights that little research exists on the use of games in writing center tutor education. What does exist tends to include games used during individual class periods (Zimmerelli 2008) but not sustained play over the course of semesters. The value and use of play pedagogy has also been underutilized in related disciplines like writing studies (Colby 2017), though that trend is changing. This chapter examines the incorporation of role-playing game (RPG) elements into tutor education to help students find their place in writing center discourses.

Gamifying elements of the course provided tutors with choices, allowing them to expand their expertise along their chosen avenues of skill level and interest. Tutors were initially apprehensive about the game aspects of the course but found that the quests helped them realistically develop as tutors because the system was flexible enough to take into account their varied academic, personal, and professional backgrounds. Taking on the role of a character for their first semester helped them see themselves on a journey where they would employ manageable strategies built from course readings and previous experiences until they could better identify as knowledgeable peer tutors capable of taking on

https://doi.org/10.7330/9781646421947.c010

a wide range of peer tutoring sessions. Overall, students felt they built a sense of autonomy, community, and tutor identity over the semester.

In designing my own quest-based course, I included both mandatory and voluntary quests that were submitted weekly throughout most of the semester.[1] Tutors used quest logs to document their completed quests and reflect on how this work shaped them as tutors. This quest-based structure draws heavily from RPGs, narrative-based games in which players create unique characters from a set of archetypes and collaborate with other players to complete narrative-based quests. These archetypes break down into specific classes or roles based on the skills characters possess that aid and complement their team. Popular examples of RPGs include *Dungeons & Dragons* and *World of Warcraft*. This RPG structure would work well in a center that employs students starting with a wide range of experiences and abilities because it gives tutors the opportunity to start approaching content at their own comfort level.

TURNING A TUTORING COURSE INTO A RPG

Local Context

In fall 2017, I started in a tenure-track position as an English professor and writing center director at a small, diverse, four-year private college. Our center is not affiliated with student success initiatives on campus, giving me freedom on how to structure our training. I also inherited a one-credit pass/fail course called Tutoring Writing: Theory/Practice. In fall 2017, I taught a discussion-based tutoring course as I acclimated myself to the college's culture. Tutors kept up with readings and talked in class, but I felt I could be doing more to encourage them to use what we were learning in class in their peer sessions and administrative work.

I knew that tutors needed opportunities to experiment. Prior to directing this center, I served as the assistant writing center director at a large public university that strictly employed graduate-level English majors. There, I learned that tutors worked effectively when afforded opportunities to develop self-paced projects within the center, even if those initiatives ultimately failed. This idea of play and risk is supported in research on writing centers. In a study by Carol Peterson Haviland and Marcy Trianosky (2006), writing center tutors described their ideal directors as organized without being overbearing or controlling (312–13). They wanted opportunities to take risks and to have faith that their directors would support them and defend their work to other faculty and administration. Essentially, tutors needed some space to safely play within their roles.

I wanted to combine previous lessons on risk-taking spaces with the idea of narrative building. I looked for ways to provide more structure and guided choices in this new center, especially as my new tutors established their own writing praxis. This structure had to be highly interactive because as Muriel Harris (2006) explains, reading about tutoring is not necessarily enough to internalize and embody a center's values and principles (304). Finally, as Haviland and Trianosky (2006) highlight, problems can develop when tutors see their directors as teachers because the teaching relationship can actively work against directors encouraging "tutors to create new knowledge rather than follow established practices or shape their new ideas into existing theoretical frames" (317). Games are a great way to establish interactivity and de-center the teacher in learning environments. For instance, Lisa Zimmerelli (2008) argues for the incorporation of serious games into her tutoring courses, noting that "they move the epicenter of knowledge from the teacher to the group" (101). With all this in mind, I created a course centered around common texts that encouraged autonomy and risk taking and helped establish a community.

In building play into tutor education, I adapted RPG elements into my tutoring course because they allowed for a degree of experimentation within a common experience. R. Mark Hall (2017) asserts that we can build stronger writing centers with a focus on common texts. While Hall focuses on texts used to run centers (e.g., intake forms and observation notes), syllabi and training materials also apply. I did this for two principal reasons that will be explored further in the next section. The first reason was that role-playing games encourage playful participation that allows for collaboration, identity construction, and failure. The second reason is that RPGs allow for narrative building, and I wanted tutors to consider their contributions to the stories surrounding centers early in their tutoring careers. Of course, as of yet, little is written on using RPGs in tutor education. Instead, I turned to the developing academic work on using RPGs to learn other skills, particularly writing.

Playful Pedagogy and RPGs as Educational Tools

The field of play pedagogy is most definitely still evolving, but many scholars have argued that play is an opportunity for students to critique the systems and greater cultural narratives they find themselves in. Play studies began, as Brian Sutton-Smith (1997) notes, with an interest in how children learn and with perhaps an overemphasis on play as a form of imitation (7). However, higher education has certainly carved out a significant space within game studies, expanding the ways games can be used to practice critical thought. Play provides an opportunity for a space to reexamine preconceived understandings about how processes (such as writing) work. James Paul Gee (2007) observes, "What video games do—better than any other medium in my view—is let people understand a world from the inside" (16). Games position players inside a system of rules, encouraging them to figure out which to follow, resist, or ignore. In these rule systems, games allow for both imitation of and resistance to entrenched narratives. Miguel Sicart (2014) calls this the carnivalesque element of play because play "appropriates events, structures, and institutions to mock them and trivialize them, or make them deadly serious" (19–20). By establishing a student or new tutor in a space where play is acceptable, educators are inviting them to consider how systems work and what their place is within them. This critical thought could be helpful to students understanding how the writing center works within the larger ecology of the college or university.

Play encourages players to break down narratives into systems of values and beliefs we can in turn find our place in. As Danielle Roach (2015) observes, "Though the framing of play as progressive seems to dominate the scholarship in this particular area, the social constructivist perspective would seek to extend the conversation into discussions of power. In this way, play evolves from simply a tool of passive socialization to a broader indicator of underlying power structures, possibly anticipating the next move (which, arguably, is to somehow win the game by better understanding the political and social rules by which it is played)" (12). Play not only invites us to collect the resources and strategies to win but also invites players to consider who holds power within any system and what they intend to do with it. Elliott Freeman (this volume) channels Ian Bogost to liken writing centers—and indeed higher education—to playgrounds. Playgrounds use loose structures and boundaries to invite children to interpret reality in any number of ways. Children create games and rich stories out of what is available to them. Games but, most important, play extend beyond ideas of fun and make available avenues for critical thinking and emotional investment.

In addition to the critical thinking affordances of play, play is also work. This work can be focused to help players enhance both their skills and their overall perception. In arguing for increased use of games in the writing classroom—and by my extension the writing center—Rebekah Shultz Colby (2017) claims that play is a way for students to do the hard work of navigating ideologies, giving examples such as having students play a character of a different gender in massively multiplayer online role-playing games (MMORPGs) to more practically understand feminist and queer theory (56–57). Games highlight the how and why of work as well. Judd Ethan Ruggill and Ken S. McAllister (2011) describe what they call "gamework," acknowledging the ways "gaming articulates work, labor, play, and pleasure together" (92). Finally, in our consent to play a game, we practice the work of entering into and challenging existing power dynamics, or as Jane Friedhoff (2016) eloquently puts it, "when we make a game about the power we need, or the world we wish we had, we're not only creating a space where we can embody that power: we're granting each other the language to talk about that world, talk about that gap, talk about that desire" (par. 29). Understanding the design of games helps players better understand the inner workings of power systems. In the context of writing centers, play can help us see issues of gender, race, and power that surround college writing and find ways to discuss and address these issues in kind.

A clear advantage of using games to educate tutors is that most games begin with learning to play the game. They give players manageable challenges at the beginning and become more challenging over time. Harris (2006) argues that we can treat tutor education in a similar fashion by structuring training around a writing center appointment, from warm greetings to meeting tutors where they are, to helping tutors command the tutoring process, and ultimately, "to play against the hierarchy that is inherent in the situation, both in the tutorials to come and during the training time" (305). Anne Ellen Geller and her colleagues (2007) call this a trickster mind-set, "one that can be awakened to and can awaken moments of discernment about uncertainty" (16). Games, when used appropriately, have the power to help tutors engage critically with course content and to build an identity within the largely textual framework of a course.

Role Playing to Engage in Writing Center Narratives

Narrative building in writing center work is critical because the daily work of writing centers varies, and most faculty and students only have limited and often distorted understandings of the work of any specific center. Jackie Grutsch McKinney's (2013) *Peripheral Visions for Writing Centers* begins with a list of twenty-five tasks a center director (and their tutors) might be handling at any point in time (1). McKinney's list (and book) consequently argues that "writing center work is complex, but the storytelling of writing center work is not" (2) and challenges center directors to question the master narratives around their centers. Writing centers are rhetorical spaces that define campus writing culture in part because tutors help guide students in writing habits, both within the context and outside of an individual course, professor, or discipline. At the same time, centers function as contested spaces that intersect with distinct visions held by students, faculty, and administrators. Tutors need to be introduced to these conflicting narratives, as they will encounter them in their day-to-day work. For example, if a professor sends a student to the center to have their paper copyedited, the student might be surprised to learn that many writing centers initially focus on higher-order concerns. Here, two identities are at play: writing center as copyediting service and writing center as a learning space to improve writing practices. In addition to these competing identities, centers are also situated in particular spaces (e.g., libraries, English departments, learning commons) that work under several layers of rules and expectations outside of best tutoring practices.

Conflicting narratives on writing centers are rarely resolved by professors and writing center directors. Directors use several professional channels (e.g., listservs, Facebook groups, Twitter chats) to help them construct and defend their unique center identities. However, center-created narratives are often based on a sense of idealism that might not be useful for every center. Those with the greatest responsibility over a center narrative are the tutors and writers who find themselves working together during a session. Essentially, we need more complicated narratives about what writing centers do; and we need ways to help tutors, students, faculty, and administrators understand these narratives. McKinney (2013) comes to this same conclusion, noting that we are already doing this unique work: "We have hardly started recognizing the scope of what is already done in the writing center. So, I don't think it is fair to stick to a narrative that does not fully represent the possibility, promise, or actuality of our work on the basis that the narrative makes our work sound legible or helpful" (80). McKinney stresses that this narrative rewriting is going to take directors researching and writing about how many different writing centers work.

There are multiple ways to help tutors navigate these discourses. However, RPGs provide rich structures that help new tutors better understand and develop their own sense of these narratives. RPGs allow for narrative interpretation in context-driven spaces. R. Allen Cheville (2016) notes that these "narrative modes strive for situatedness and connection to others" (808). Tutors must connect with the writers who use the center, with professors who assign writing in their classes, and with the places these centers inhabit. Tutor education is the best place to introduce tutors to these many discourses.

Overlaying an RPG onto a Tutoring Course

I confess that I have spent years attending panels at academic conferences and listened cynically as colleagues discuss the incorporation of RPG mechanics into their courses. As an RPG tabletop player, the biggest issue I had was that I knew how much time RPGs can take. I was not willing to give up my course meetings to campaigning, especially because the course I taught was limited to one hour per week.[2] However, at the 2018 Computers and Writing Conference, Cody Reimer presented on using choice to encourage participation in the classroom. Reimer suggested using decision trees (decisions made during play that can alter the game's narrative, alter a character's skills, or both) to give students more autonomy over their work. In games, players often earn experience points as they play. The player can use these points to improve their character, but they often have to make choices about these improvements. A player might put all their points into becoming stronger or sneakier or learning magic spells. They might invest in a little of everything, knowing they will be kind of good at everything but the master of nothing. In Reimer's example, these choices specifically allowed students to determine what projects they completed to show course content proficiency. Instead of setting up the class as a tabletop RPG, this suggestion promoted the use of play-by-post RPGs (PRPGs), which allow players to self-pace their progress in a persistent world. Play-by-post games can follow the same rule books as other RPGs, but players engage asynchronously, either online (forums or email) or through the mail. This stretches out the length of each campaign but often allows players more time to thoughtfully build a story. Borrowing Reimer's structure, I designed simple self-guided, quest-driven tutor training activities. Tutors chose two of three RPG class roles to dedicate themselves to during this course, and those roles helped the tutors build specific identities. The intention was that these class roles would give them an identity they could freely play with to examine their relationship to the many narratives that surround writing center work. Ultimately, I chose the class roles of summoner, paladin, and cleric. I decided on these classes because they allowed for a variety of activities and outlooks and avoided more combat-based characters. No warriors or barbarians because tutors are not at war with students, faculty, or the administration.

Summoner quests focused on field and institutional research. The goal was to have multiple opportunities for tutors to investigate the narratives that surround writing center communities outside our institution. In addition, past tutors had gone on to graduate programs based on research done in the center, and I imagined that this opportunity would help new tutors see themselves as both practitioners and researchers. Some example quests included students researching similar writing centers, writing center scholarship, and common genres used by our writers. They could also attend additional professional development.

Paladin quests focused on advocating for and promoting the center on campus. Tutors created documents that required them to articulate their work in the center to outsiders. Some sample quests included recording and reflecting on their own sessions, doing classroom visits, making memes for social media accounts, and making posters to advertise the center.

Cleric quests focused on improving the well-being of the tutoring team. They were intended to build a community of tutors who relied on each other. Sample quests included helping with a write-in, making new handouts, proofreading materials, and picking up shifts for other tutors.

Each RPG class role came with a list of mandatory and optional quests; tutors had to complete ten in any order they chose (appendix 10.A). Tutors kept track of their quests in a digital quest log. I checked logs every week and provided written feedback that was intended as a conversation and encouragement to develop their observations, perhaps in a follow-up quest. I did not comment in depth on the products they created because, as Zimmerelli (2008) observes, the process in playful pedagogy is often more important that the product (101). In our course meetings, we discussed the quests in addition to covering course readings and questions they had about tutoring. At the end of the semester, tutors had to complete a final project, which was to create something that would improve our center. Examples of final projects have included making an exit survey for writers, developing a new standard template for our help docs, and developing activities to help writers transfer skills from one course to the next. As such, quest logs augmented course readings and gave tutors introductions to many elements of writing center work, helping them make larger contributions to the center through their final projects.

THE EFFICACY OF A QUESTING COURSE

I wanted to see what impact this quest-driven course had on tutors. After the completion of the fall 2018 semester, I emailed an Institutional

Review Board (IRB)–approved voluntary survey (appendix 10.B) on quest logs and asked for permission to code tutors' quest logs for major patterns and themes. The survey consisted of twenty-one questions adapted from a learning experience questionnaire developed by KTH Royal Institute of Technology (Borglund et al. 2016). The majority of these questions asked tutors about the meaningfulness and manageability of quests on a five-point Likert scale. Tutors also completed five short-answer questions, giving them the opportunity to discuss specific quests or improvements for the project. This case study is small, but small case studies often provide useful information and are common in the writing center field because tutors at any one school are a limited sample set.

Of the six tutors in the course, five completed the survey, answering all questions and permitting a review of their quest logs. Tutors were positive or neutral about each course element included in the survey. From their responses and my own reflections, I wanted to discuss three common themes addressed in their survey feedback and quest logs. First, the quests gave tutors a sense of autonomy and control over their own learning environment. That being said, tutors actively sought out ways to collaborate, mentor, and build community inside this somewhat self-guided structure. Finally, tutors spent a significant amount of time on identity construction, with the majority of tutors identifying their roles as tutors within the larger campus community.

Autonomy

Tutors expressed that they had a great deal of autonomy, particularly with the flexibility they had in choosing quests. As one tutor noted, "The best aspect of the mini quests was the ability to pick and choose different levels of assignments every week. Some weeks, I could relax and do something fun and manageable, but some weeks I could challenge myself to [do] something out of my comfort zone." Another tutor enjoyed being "able to pick which mini quest you wanted to do based on your workload for the week. I found that it helped me enjoy completing them, because I knew that I was able to save the bigger quests for a week where I had ample time to do them." With that flexibility, tutors expressed that they were able to explore content on their own, which can help tutors extend themselves at their own pace and give them ample space to practice a critical understanding of the center. Being able to pace their choices also meant that tutors expressed being confident that they could complete quests with their current background knowledge. While they may have not felt that way at the beginning of the semester, the quests allowed them to slowly build that knowledge. In their short responses, one tutor did suggest getting rid of the class roles altogether and allowing tutors to choose any optional quests. One reason the student may have suggested this is because some quests were out of my control. For instance, in the summoner class role, one option was to attend North Carolina's Tutor Training Day, which was then canceled and left summoner tutors with one fewer option. However, we discussed this in class, and tutors mentioned that the class roles forced them to try new things they would have avoided with more options, so in the future I will simply have more choices for each class role.

An examination of choice is important here. While students had many choices, they made similar ones. All five tutors in this study chose the paladin class (which dealt with promoting the center), three chose summoner (dealing with researching writing centers), and two chose cleric (helping other members of the center, including those outside the class). Students picked two class roles, so ten class roles are included. Initially, there were four mandatory quests (make a personal writing center appointment, annotate an academic article on writing centers, record an appointment, and help with the write-in). However, after meeting with veteran tutors, the write-in had to be scheduled after most new tutors had finished their quest logs. The three other mandatory quests were completed by all tutors and were essential to helping them understand the relationships and power dynamics of writing center appointments. Of the mandatory quests, most tutors elected to either make an appointment or annotate an article first. All students recorded an appointment as either their last or next-to-last quest. Selecting to finish the semester with the recording meant that tutors had more experience and were able to ask a writer they had previously worked with.

Of the optional quests, the most popular were find three memes, make a meme for our Instagram account, reflect on a passage from McKinney's text, pick up a shift, make a promotional poster, propose a workshop, and review our website. The popularity of making memes may have driven all participants to choose the paladin class. However, as I discuss in the identity section, tutors may also have most closely identified as members of the college over being tutors or researchers this early in their tutoring careers. Some examples of quests not chosen by any tutor include create a video to promote the center, join a writing center listserv, do some copyediting work for the center, and propose a redesign of our center. Several of these unselected quests required more guidance to complete, and at least two of them would require more than one week to do well. It may be that these kinds of quests do not lend themselves to the current quest format or, as will be discussed later, the quests need to be written to promote collaboration between tutors.

After looking at some of the lower-scoring questions on this study, the majority of tutors reported that they felt neutral about whether the quest log offered an opportunity for reflection. Tutors did receive regular feedback and were able to turn an earlier quest into a final project. For instance, if they proposed a workshop on Chicago style in their quest logs, then the final project could be making the outline and handout for that workshop. All of the final projects were based on something they discovered about the center after working on their quests. In my initial design, the quest logs were the reflection, but I never said that explicitly and will do so in the future to help students recognize quest logs as reflection.

Community Building

Tutors reported using quests as a way to build community in the center. This included building director/tutor communication, building mentor relationships, and collaborating on quests. Quest logs allowed me to give written feedback followed by verbal feedback during course meetings. I was able to keep in regular contact with students even if I couldn't check in with them during their shifts. One surprise for me was that tutors responded positively to having the chance to collaborate. While quests were intended to be responded to individually, I was happy to find that tutors turned to other tutors to help them complete projects, including the help of veteran tutors who experienced a different course structure. This collaboration helped establish a foundation for a community within the center. Part of the collaborative feel may have come from pulling narrative elements from collaborative RPGs to structure the quests. With that consideration, even though this first iteration of the quests did not privilege collaboration, the tutors did privilege that based on how they played.

When asked about their favorite quests and why, tutors noted:

- "Helping with the write-in."
- "The article annotation was my favorite one. It helped me practice annotating for future reference, and I got to read an interesting journal article."
- "I really loved creating my own meme because I love memes and expressing myself."
- "I really enjoyed the quest that involved us recording one of our sessions and reflecting on our strengths as well as areas in which we could improve."

- "Making a poster for the Writing Center bit made me feel like I had made a good contribution."

A few generalizations can be pulled from these favorites: they can be done individually and in the student's own time, they can allow for a good bit of freedom and creativity, and tutors see a direct correlation between the work put in and the benefit to themselves or the center. For instance, write-ins are group events sponsored by the center that provide refreshments and tutors present to answer questions. No one-on-one appointments are scheduled at these times. While write-ins are communal center-wide events, tutors tend to have different levels of engagement, and they come and go as their schedules allow. However, tutors enjoyed being together to host the event and invite their friends into their work. Suggestions for changing the class included adding quests that dealt with grammar and citation styles and more quests that dealt with community and social media building. As such, while they liked the self-paced quests, they liked quests that involved interacting with or helping others.

Identity Construction

It was in building tutor identities that the most narrative building could be seen. The quest structure admittedly asks tutors to take on an identity from the beginning of the semester with little knowledge of writing center tutoring. As in a traditional RPG, identities revolve around what one does to help their team. Archetype work is important in story building because it also allows tutors to take on a professional identity and imagine themselves as tutors before they feel ready to embody the role themselves. As such, while archetypes are mostly about what a player can do, they project a person the player can aspire to be. Being a person who helps their peers, their college, or their field is a pretty powerful place to start any semester.

As mentioned above, most tutors chose the paladin class role. I confess that I imagined that cleric would be the most popular of the class roles. I hypothesized that students who had willingly signed up for a job that centered on helping other students would be excited about a class centered on helping their co-workers. This bias may be tied to my own identification as the director (and champion for all tutors and writers who visit our center). However, the fact that every tutor selected the paladin class perhaps hints at a different understanding of what students are often best at: getting excited about things and sharing that excitement with others. I often find that promoting the center is my least favorite job. Maintaining social media accounts, coming up with new posters, and sending out emails can all feel like a waste of time. I would rather spend that time researching writing centers and developing tutors. Our tutors, none of whom currently have any intentions of going on to do graduate work in English, were mostly interested in understanding their tutor identities in relation to the college. As seen above, they enjoyed making posters and memes. They liked seeing their work up in hallways and on Instagram. The paladin class, as stated above, also had the most quests where students were able to see a clear connection to the application of their work.

As paladins, the recorded sessions were the quest that most cemented their identities as tutors and helped them see that improving as tutors is a constant process. Hall (2017) discusses having students record and analyze a sample session and develop "a heuristic for working together with writers to learn the context for writing" (47). Of the recorded sessions, one tutor reflected, "I want to say that this particular assignment was unnecessary torture, but it was actually very necessary torture. :(" The student went on to mention the things they noticed about their own approach and what they would do in the future to improve. This quest forced the student to reconsider the narrative they had imagined about their own tutoring. After these recorded sessions, students made individual appointments with me that allowed us to talk through the sessions. All of these appointments were productive and worked well at the end of the semester (as tutors typically finished the semester with this quest). Ultimately, the quest and follow-up meeting were important to help students identify as tutors, and the quests had prepared them to critically analyze their tutor identity and make specific changes.

Director Perspective

After reviewing student surveys and their quest logs, a few observations can be drawn from this study. Overall, a self-regulated and quest-driven structure to tutor education can seem overwhelming to new tutors, especially those who do not identify as gamers and have had little exposure to play pedagogy. However, students seemed to comprehend the course expectations and reason behind quests within the first two weeks of completing quest logs. We spent half of our first meeting going over the quests and had time at the beginning of each class to discuss and negotiate the expectations for the quests. I will say that working through this idea felt very much like an ongoing RPG campaign, where players can seek advice and negotiate with the Game Master (the director, in this instance). This ongoing conversation was itself a community-building element. Once tutors understood the project, they enjoyed it and valued the range of choices they had and the allowance this structure gave them to better fit tutor education into their already busy schedules. The game of the course was flexible enough to fit easily into the *Tetris* game that is many a tutor's busy life.

CONSIDERATIONS FOR DIRECTORS CONSIDERING QUEST-BASED TUTOR EDUCATION

Ethical Drawbacks to Play-Based Pedagogy

The use of games in the classroom is not without drawbacks. One issue to consider when using games in courses is identification. For instance, women are less likely than men to identify as gamers (Colby 2017), and gaming, particularly adopting language from niche tabletop RPGs, might make students feel at a distance from the content. Jessica Clements (this volume) uses that lack of familiarity with games, encouraging tutors to use MMORPGs as ways to explore the intersections of identity and content familiarity. Likewise, while games can call attention to embedded power dynamics, they can also easily reproduce them. Many games rely on outdated and misinformed notions of sex, gender, class, and race. In particular, Antero Garcia (2016) criticizes the game manuals[3] of the most famous RPG, *Dungeons & Dragons*, for the sustained undermining or erasure of women, people of color,[4] people with disabilities, and those identifying as LGBTQ. Finally, the course I designed uses quests and adopts what can be called gamification (the process of using game elements to encourage certain kinds of work), a sometimes maligned practice that can be used to create overly simplistic games. However, Wendi Sierra and Kyle Stedman (2012) write about the *C's the Day* game within composition studies, stating that thoughtful gamification added to their experience of the conference: "In our cases, that addition came in terms of agency and community, but there's no reason that the added value of gamification needs to stop there" (par. 14). Sierra (2013) encourages teachers to make sure they thoroughly understand the mechanics of the game they hope to use and have a clear understanding of how the learning outcomes for any class align with those of game play. Instead of avoiding the adaptation of gaming styles that carry specific cultural weight, teachers should be more careful to build games that avoid these trappings. This means taking the best of their mechanics to create games that promote critical thought.

Time

Before beginning this project, I had played through quest-driven activities—*C's the Day* at the Conference on College Composition and Communication specifically. I had also designed conference-specific quest games while I was a graduate student. These experiences made me much more willing to try this project. That being said, planning a semester that was balanced between low-stakes fun tasks and more demanding tasks was quite the challenge, and I confess that I spent as much or more time planning this one-credit course as I did three-credit courses I teach in my department. Fun almost always takes more work in any learning environment, particularly for the person designing the activities.

However, once all quests were set up before the semester started, much of the rest of the course ran smoothly. A few of the quests asked tutors to find or create promotional materials for the center (memes, digital flyers), and having these documents coming in at a staggered pace helped me have new content to put on social media and bulletin boards regularly. Essentially, tutors were able to use a variety of texts to explore the writing center's identity, and I was able to use those texts to provide our campus with focused advertising. By using a game structure, directors will also have a set and reusable system to educate tutors across semesters. This constancy is caused both because tutor education extends far beyond any formal system and because new tutors join the center every year/semester as previous tutors graduate.

The Manual

Designing clear and concise directions is perhaps the most important design feature for this kind of tutor education. I say this as a professor who also teaches professional and technical writing but still struggled with clear directions. My syllabus with quests was ten pages long. I made the mistake of calling quest logs "journals" on the first page of the syllabus. This led to much confusion because some tutors thought they had to do both quest logs and journal entries. Typically, we do all our hiring of fall tutors in the spring, but we had a tutor join us in week three of the course, and small mistakes in terminology made that catch-up process more of a challenge. In hindsight, I would have had a current tutor review the new syllabus with me prior to teaching the course.

The Center

The use of quest logs required me to provide tutors with a lot of timely feedback on their logs. I wanted to make sure I was validating their experiences and pushing them to apply even more critical thinking to the center. This feedback meant I was in regular conversation with new tutors, even those who worked hours when I could not possibly be on campus. One advantage to this ongoing written communication is that now in the spring semester, with no tutoring course, tutors continue to solicit that feedback in higher numbers than the prior year. This is especially important because due to extensive renovations of the campus, the writing center is currently in a different building than my office.

Because quests forced them to, tutors, I found, better balanced reflections on their own tutoring with considerations of how tutoring fit into campus culture. Quests that asked tutors to be critical of elements of our center (our website) also seemed to help them feel as though they had more ownership of the center. Likewise, they were able to gently nudge me on pesky low-level tasks. For example, it is much harder to put off contacting public relations about a website change after three tutors mentioned in their quest logs that the "make an appointment" button should be moved up on our home page.

CONCLUSION

Tutor education is the way we introduce students to complex writing center discourses and help tutors situate themselves into long-standing narratives. Helping orient students and guiding them through finding their places within these conversations can be overwhelming at first. Tutor education is a uniquely applied kind of learning wherein most students are tasked with tutoring while they are still learning. At the same time, most of our tutors will not go on to research or study writing centers. While directors are responsible for ensuring that tutors are good at their job, it might be even more important to tutors themselves to help them build growth mind-sets and become critical thinkers. Tutors should be looking for application and transfer, and games are great at teaching these skills.

In developing this quest structure, I hoped to find a way to introduce students to some of the many avenues of practice and research that writing center work has to offer. Overall, this was a successful pilot study. Over the course of the semester, tutors were able to adopt tutor identities they could develop into. They worked to build autonomy and community, building the role of tutor into their campus identity. They took chances and followed up with ideas. Finally, they were more likely to experiment to try to make our center a better place. Directors like Zimmerelli have shown that games can be used in the classroom to help students analyze and reflect on their identities and practices in the center. This study shows that sustained play can allow tutors more ownership of their own development.

Research into the intersections between writing centers and play is nascent at best. There's not enough data on the adaptation of playfulness to writing center spaces, and that limits the conversations the field is having now. At the same time, as McKinney says about the work of creating center narratives, the key elements already exist. For example, all the quests I used in my tutoring course could be adopted in another center without calling them a game, as they are very much work. Yet that would strip away perhaps the most important part of the whole project: the freedom to use archetypes to play with an identity. This kind of play can enrich experience, encourage analysis, and prompt serious action. As such, more attention needs to be paid to playful approaches to tutor education, and more studies need to be shared on the development of playful center models.

NOTES

1. Because this chapter discusses RPG character classes and college tutoring classes, for disambiguation I will refer to tutor education as a course.
2. In RPGs, campaigning is completing a story arc over multiple game sessions. Typically, campaigns are played with the same characters, who will grow and change. In *Dialect*, an RPG I sometimes play in class, students build characters for an isolated society and in the campaign trace the creation and destruction of that society's language over the course of three class sessions. Along the way, their characters change by creating alliances, betraying their society, martyring themselves, or deciding to become hermits.
3. Because *Dungeons & Dragons* is a narrative game, players can easily change the rules. For instance, players could decide to remove mechanics that reduced female characters' strength in earlier editions of the game.
4. Race here means both human individuals from distinct ethnic groups and demihuman characters (e.g., dwarves, elves, halflings).

APPENDIX 10.A

QUEST ASSIGNMENTS

In weeks 3–13 (Aug. 28–Nov. 13), you will maintain a quest log that reflects on your quest to learn and grow as a tutor. Most entries should be approximately 200–500 words. These quests should be done primarily while you have downtime at the Writing Center (before you do work for other classes). You will complete ten quests in all. There are four mandatory quests that give you the foundations of the game. You will also pick two classes (Cleric, Summoner, or Paladin) and choose six optional quests from these classes. Each quest can be completed only once. You may only work on one quest per week. **Quest Log Entries should be kept on one Word/Google Doc and emailed to jxxxxx@xxxx.edu before 6 am the day you come to class.**

CLERIC

Clerics feed health and boosts to the other members of the team so they can make their best plays. They tend to hang back, watch, and move where needed most.

With regards to the Writing Center, the cleric is the person who provides help for not only students who come into the center but also other tutors and maybe even the director. Are you the kind of person who brings an extra pen and paper to class in case someone else needs it? This might be the class for you.

Mandatory Quests

Help with a Write-In (by either attending or promoting the event). Write a 200-word reflection on how you helped, what went well, and how we could improve these events in the future.

Choice Quests

Design a handout that will help students to understand a common problem you see in your appointments. This quest probably won't be 500 words and that's okay. As the semester wears on, we will develop a template so that handouts look uniform.

Suggest a speaker for a staff meeting. It's great to have other people come in and talk to us about writing, genres, and ways we can help. Let me know who you think would be great to come in and talk to us. I will try to get them to come visit.

Offer to copy edit documents for the Writing Center. The director will definitely find you something. You have to ask at least 48 hours before the journal is due for something to work on.

Take a shift for a tutor who needs it. Reflect on whether or not there are any differences working at a different time of day.

Propose (with a graphic) an ideal physical layout for a writing center. The one good thing about this weird year is we may have a bit of say about changing the center next year to look and feel like a great space.

Propose some changes or additions to the mini quests. This is a first run. I'd like to add another class and/or more quest choices.

SUMMONER

Summoners are resource masters. They are high-level magicians who have spent years in the stacks learning spells that help them to commune with and summon minions to do their bidding.

In the context of the writing center, this means collecting resources, curating knowledge and lore, and working to help new generations to find their own expertise. Those deciding to become summoners will focus largely on writing center research. Do you often have ideas on how something can be run better? Have you ever lost a chunk of time doing "research" on the internet? You might enjoy the summoner class.

Mandatory Quest

Annotate 1 article from either WLN: A Journal of Writing Center Scholarship or Praxis: A Writing Center Journal. Annotations include:

- A bibliographic reference in either MLA or APA.
- A paragraph (3–6 sentences) summarizing the article.
- A paragraph (3–6 sentences) explaining if and/or how the information can be applied to our writing center.
- I will post your annotation (potentially corrected) to the Big List of Annotations. In fact, you should probably check the list before you read so that you don't annotate something that has already been done.

Choice Quests

Find three writing-related memes that could be shared on our social media accounts. Describe how this meme represents our center and when we should share it.

Join the WC-Listserv. Write about one of the more interesting discussions going on that week.

Attend the North Carolina Tutor Training Day (details early in the fall semester). Write about your experience meeting with tutors from other colleges.

Review the Writing Center Website. Propose a few suggestions that might have more students using the site and finding the information they need.

Propose an in-class workshop that we could offer. Maybe we could help a class identify the differences between APA and MLA or we could talk to a class about the value of peer review and/or revision.

Analyze a genre you are unfamiliar with (500 words—you probably won't be able to answer every question). Channeling Irene Clark's [1999] "Addressing Genre in the Writing Center," I want you to:

Select a genre you are unfamiliar with.
Collect at least three authentic samples of the genre. You might choose the sample of the annotated bibliography, then collect samples of annotated bibliographies from several communities that use them.
Look at the samples and describe the situation and work of the genre.

Context: Where does the genre appear? In what discipline?
Subject: What problem, issues, or questions does the genre address?
Work of genre: Why do writers write this genre and why do readers read it?
Participants: Who uses this genre? Who writes this genre? Who reads texts in this genre?

Identify and describe the recurrent features of structure and language in the genre.
What context is typically included? How does the author situate their work with other writers? What is left out?
What rhetorical appeals are appropriate?
How are texts within the genre structured and organized?
In what format are texts of this genre presented? What layout is common?
What diction (types of words) is most common? Is jargon used? Is the language formal or informal?

PALADIN

Paladins have a lot of hit points and typically charge into battle, drawing the most attention (and the most fire), allowing other members to work more surreptitiously. They believe strongly in the cause that they fight for and put themselves out there for the team.

In the writing center, this means being a spokesperson for the center. You're happy to be the face of "writing tutors" and want to do things that draw attention to the center. Getting students in the door and helping them learn what we do is a primary objective to the paladin. Have you ever unironically said "Charge!"? Consider a future as a paladin.

Mandatory Quests

Make a personal writing center appointment for something you are working on. Have the tutor work through the piece with you. Write a 250-word reflection on what it was like to be tutored. What did you notice about your own work? What worked for you as someone being tutored?

Record and reflect one of your appointments (as with the student's permission). Write a 250-word reflection that highlights one thing you never noticed about your sessions, three things you did well, and one thing you would like to work on in the future.

You must obtain written permission from the writer before you record. You must turn in a completed consent form, including both the client's signature and yours.
You must transcribe at least five minutes of the transcript and talk about it.
You will need to share the audio file with me.
One week after I receive this journal entry, we will sit down and talk through your appointment privately.

Choice Quests

Create a meme that could be shared on our social media accounts. Describe how this meme represents our center and when we should share it.

Do a WC class visit (you will also be paid for a 30 minute shift). Visit a class and give them information about the writing center. Be sure to ask questions. Write a 200-word reflection on the visit, what went well, and what to change in the future.

Create a poster advertising the center. Great posters will be posted around campus to remind students of the center.

Create a Halloween-themed video advertising the center. Must be finished before October 15th. You can collaborate with another tutor.

Design a handout that tutors could share that focuses on a common topic you observe in appointments. This can be inspired by an existing handout, but it must be adapted by you.

Reflect on the following passage from *Peripheral Visions of the Writing Center*. McKinney ultimately argues that this kind of stereotype might be hurtful to centers. Why does she think that?

> *If a scriptwriter went to a writing center conference or read much writing center scholarship, he or she might well come up with a character who was smart yet insolent. Someone who went to a college but was not part of the college—someone who doesn't know the name of the football team's quarterback, doesn't own a hoodie or bumper sticker with the school logo, but one who is on a first-name basis with the reference librarian, runs the student Greenpeace chapter, and whose best friend is the town's record-shop owner. There is a type of student who seems to find his or her way into writing centers as a tutor and a type of professional who is drawn to writing center administration, and I don't think this is accidental. Throughout writing center scholarship is an ongoing notion that writing center work is different, non-traditional—iconoclastic—and thus those who work there are, too. Thus, the image of a "typical" writing center tutor or administrator can become a visual habit,*

> *something we are accustomed to seeing and looking for and can lock us in a pattern of expectation and fulfillment.*
>
> –Jackie Grutsch McKinney, *Peripheral Visions for Writing Centers* ([2013,] p. 35).

Analyze what patterns are significant to the community that uses them: what values, assumptions, beliefs, and goals are revealed through your analysis?

What do participants have to know or believe to understand or appreciate the genre?
Who is invited into the genre and who is excluded?
What roles for writers and readers does the genre encourage or discourage?
How is the subject of the genre treated? Do they use first person or third? Why?
What attitude towards the reader is implied?
What attitude towards the world is implied?

You will not be able to answer all these questions in 500 words. Pick and choose the most important.

APPENDIX 10.B

SURVEY QUESTIONS

How many hours per week on average did you work on the course (including scheduled class time)? [0–2, 3–5, 6–8, 9+]

Meaningfulness (5-point Likert Scale)

I worked with interesting issues.

I explored parts of the subject on my own.

I was able to learn on my own by trying out new ideas.

Comprehensibility (5-point Likert Scale)

I understood how the assignment was organized and what I was supposed to do.

Understanding of key concepts had high priority in the mini quests.

The mini quests helped me to achieve the intended learning outcomes efficiently.

I received regular feedback that helped me to see my progress.

Manageability (5-point Likert Scale)

My background knowledge was sufficient to complete mini quests.

I regularly spent time to reflect on what I learned.

I was able to learn in a way that suited me.

I had opportunities to choose what to do.

I was able to learn by collaborating and discussing with others.

General Short Responses (short answer)

What was the best aspect of the mini quest assignment?

What was your favorite quest?

What would you suggest to improve?

What advice would you like to give to future participants?

Is there anything else you would like to add?

REFERENCES

Borglund, Dan, Ulf Carlsson, Massimiliano Colarieti Tosti, Hans Havtun, Niclas Hjelm, and Ida Naimi-Akbar. 2016. "Learning Experience Questionnaire—Course Analysis for Development." *Kungliga Tekniska Högskolan* (KTH). https://intra.kth.se/polopoly_fs/1.661155!/LEQ_Guide_v2.pdf.

Cheville, R. Allen. 2016. "Linking Capabilities to Functionings: Adapting Narrative Forms from Role-Playing Games to Education." *Higher Education* 71: 805–18.

Clark, Irene. 1999. "Addressing Genre in the Writing Center." *Writing Center Journal* 20, no. 1: 7–32.

Colby, Rebekah Shultz. 2017. "Game-Based Pedagogy in the Writing Classroom." *Computers and Composition* 43: 55–72.

Friedhoff, Jane. 2016. "Playing with Resistance." *Medium.* https://medium.com/@jfriedhoff/playing-with-resistance-a483b19d4fe7.

Garcia, Antero. 2017. "Privilege, Power, and Dungeons & Dragons: How Systems Shape Racial and Gender Identities in Tabletop Role-Playing Games." *Mind, Culture, and Activity* 24, no. 3: 232–46.

Gee, James Paul. 2007. *Good Video Games + Good Learning: Collected Essays on Video Games, Learning, and Literacy.* New York: Peter Lang.

Geller, Anne Ellen, Michelle Eodice, Frankie Condon, Meg Carroll, and Elizabeth H. Boquet. 2007. *The Everyday Writing Center.* Logan: Utah State University Press.

Hall, R. Mark. 2017. *Around the Texts of Writing Center Work: An Inquiry-Based Approach to Tutor Education.* Boulder: University Press of Colorado.

Harris, Muriel. 2006. "Using Tutorial Principles to Train Tutors: Practicing Our Praxis." In *The Writing Center Director's Resource Book,* edited by Christina Murphy and Byron L. Stay, 301–9. Mahwah, NJ: Lawrence Erlbaum.

Haviland, Carol Peterson, and Marcy Trianosky. 2006. "Tutors Speak: 'What Do We Want from Our Writing Center Directors?'" In *The Writing Center Director's Resource Book,* edited by Christina Murphy and Byron L. Stay, 311–20. Mahwah, NJ: Lawrence Erlbaum.

Hymes, Kathryn, and Hakan Seyalioglu. 2019. *Dialect.* Washington, DC: Thorny Games.

Johnson, Karen Gabrielle, and Ted Roggenbuck, eds. 2019. "How We Teach Writing Tutors." *WLN: A Journal of Writing Center Scholarship.* https://wlnjournal.org/digitaleditedcollection1/index.html.

McKinney, Jackie Grutsch. 2013. *Peripheral Visions for Writing Centers.* Boulder: University Press of Colorado.

Reimer, Cody. 2018. "Flip It, Game It, Play It, Grade It: A Harder, Better, Faster, Stronger Pedagogy for FYC." Paper presented at the Computers and Writing Conference, Fairfax, VA. May 24–27.

Roach, Danielle. 2015. "Pedagogy at Play: Gamification and Gameful Design in the 21st-Century Writing Classroom." PhD dissertation, Old Dominion University, Norfolk, VA.

Ruggill, Judd Ethan, and Ken S. McAllister. 2011. *Gaming Matters: Art, Science, Magic, and the Computer Game Medium.* Tuscaloosa: University of Alabama Press.

Sicart, Miguel. 2014. *Play Matters.* Cambridge, MA: MIT Press.

Sierra, Wendi. 2013. "Gamification as Twenty-First-Century Education." PhD dissertation, North Carolina State University, Raleigh.

Sierra, Wendi, and Kyle Stedman. 2012. "Ode to the Sparklepony: Gamification in Action." *Kairos: A Journal of Rhetoric, Technology, and Pedagogy* 16, no. 2. http://kairos.technorhetoric.net/16.2/disputatio/sierra-stedman/tng.html.

Sutton-Smith, Brian. 1997. *The Ambiguity of Play.* Cambridge, MA: Harvard University Press.

Zimmerelli, Lisa. 2008. "A Play about Play: Tutor Training for the Bored and Serious, in Three Acts." In *Creative Approaches to Writing Center Work*, edited by Kevin Dvorak and Shanti Bruce, 97–114. Creskill, NJ: Hampton.

PART 3

Staff and Writing Center Education Games

11

WRITING CENTER SNAKES AND LADDERS

Nathalie Singh-Corcoran and Holly Ryan

Objectives

- To productively explore discomfort in the writing center.
- To offer an opportunity for lateral mentoring.
- To support discussion of difficult tutoring situations.
- To provide space for reflection and planning for upcoming difficult tutoring scenarios.

Number of Players

Two to four players.

Materials

- One *Snakes and Ladders* board (templates can be found online). The numbered squares on the board should go to 100.
- Two to four small pieces of candy from a variety pack (e.g., Hershey's Miniatures, Starburst candies).
- Thirty "scenario" cards (see supplemental materials section).
- One six-sided die.

Starting the Game/Setup

1. Each player chooses a different piece of candy to serve as their playing piece.
2. Each player places their piece off the board, just before square 1.
3. Players roll the die to see who goes first. The person who rolls the highest number goes first.

Rules of Play

1. To begin playing, the first player rolls the die and moves their piece according to the number they rolled. For example, if a player rolls a 3, they move their piece three spaces. Each subsequent player then takes their turn.

https://doi.org/10.7330/9781646421947.c011

2. If a player's piece lands on a ladder, they "climb up the ladder" and move to their new position on the board.
3. If a player's piece lands on a snake, they "slide down the snake," move to their new position on the board, AND draw a scenario card.
4. If a player draws a scenario card, they read it aloud, discuss how they would handle the scenario on the card, and then ask for feedback from the other players. A 5- to 10-minute conversation about the scenario should ensue. Some scenario cards are blank. If a player draws a blank card, they are to draw from their own experience and discuss a real-life and recent writing center scenario.

Winning/Completing the Game

Play continues until the first person reaches 100 and wins the game.

Variations

Players can continue playing the game to see who wins second place, third place, and so on. Game play can continue as long as scenario cards are available.

Rationale for the Game

In 2017, the *Peer Review* published a special issue on writing centers as brave/r spaces. In the introduction to the special issue, Rebecca Hallman Martini and Travis Webster (2017) make the following call for social justice in the writing center: "The concept of writing center as brave*r* space recognizes the discomfort and challenges involved when tutors, students, faculty, and staff engage in everyday conversations about privilege and difference. Rather than considering this kind of work to be outside of our writing center purview, we argue the exact opposite; brav*er* work is not only crucial to writing center work, but we are also well positioned for brave acts" (original emphasis).

Hallman Martini and Webster highlight that tutoring is not apolitical and can be uncomfortable. Centers cannot be thought of as safe spaces but rather as spaces where tutors need to confront difficult texts, ideas, writers, and systems. The rest of their special issue defines, explores, problematizes, and operationalizes what it means to be a brave/r writing center.

To educate tutors about how to work in a brave/r writing center, tutors need readings and active learning opportunities to learn, practice, and reflect on the tenets of this style of tutoring. They also need to learn to live in the discomfort of difficult sessions. Shannon McKeehen (2017) provides some suggestions for how tutors can learn critical empathy and collaborative fact finding as a way to challenge misinformation in a writer's paper. Questioning strategies, applying discourse analysis, and drawing on conflict resolution tips are all ways tutors can tutor productively in an uncomfortable situation.

In addition to tutoring strategies, tutors need opportunities in their tutor education to discuss, reflect, and mentor each other on ways to productively deal with discomfort. In the West Virginia

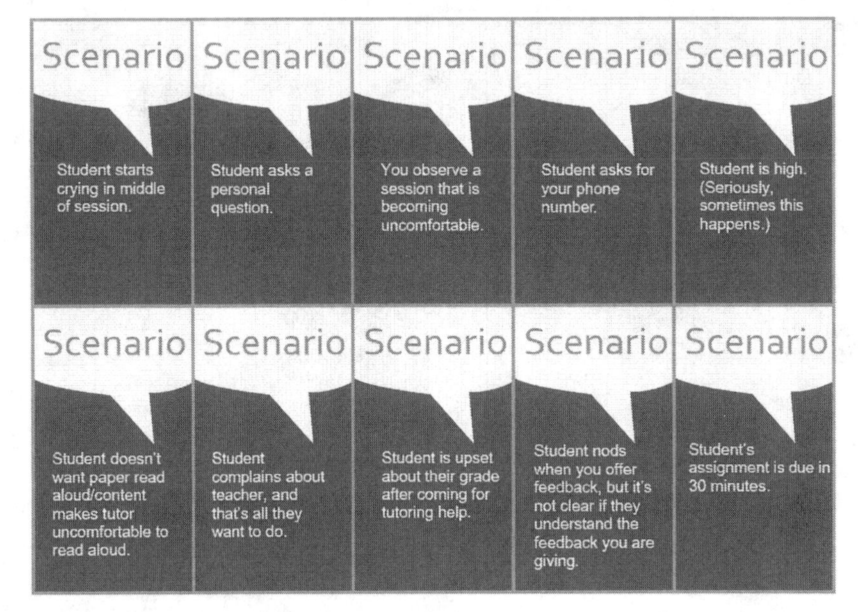

Figure 11.1. Scenario cards template

University writing center, we use a writing center version of the board game *Snakes and Ladders* as a means to enhance our thinking about dissonance in tutoring. *Snakes and Ladders* is used as a tool not only to illustrate the highs and lows of tutoring sessions but also as a means to imagine how to be brave/r in difficult situations.

REFERENCES

Hallman Martini, Rebecca, and Travis Webster. 2017. "Writing Centers as Brave/r Spaces: A Special Issue Introduction." *Peer Review* 1, no. 2. http://thepeerreview-iwca.org/issues/braver-spaces/writing-centers-as-braver-spaces-a-special-issue-introduction/.

McKeehen, Stacey. 2017. "Critical Empathy and Collaborative Fact-Engagement in the Trump Age: A Writing Center Approach." *Peer Review* 1, no. 2. http://thepeerreview-iwca.org/issues/braver-spaces/critical-empathy-and-collaborative-fact-engagement-in-the-trump-age-a-writing-center-approach/.

SUPPLEMENTAL MATERIALS

Numerous *Snakes and Ladders* templates are available on the web, and there's a Pinterest board with examples. Pro tip: Consider using a readymade *Chutes and Ladders* board.

Scenario cards template (see figure 11.1).

12

ACTIVE LISTENING UNO

Stacey Hoffer

Objectives

- To engage faculty and staff in learning vocabulary for active listening.
- To help conceptualize active listening and provide a foundation for discussion.

Number of Players

A minimum of four players.

Materials

- Index cards.
- Marker.

Starting the Game/Setup

1. Create the deck.

   Using index cards and a marker, create three different types of cards.

   a. The first category: terms used to describe active listening.
   b. The second category: explanations or definitions of the terms.
   c. The third category: examples of the terms.
2. Create the "rule" cards.

   Using index cards and a marker, create at minimum two cards per "rule." The rules are: star, skip, reverse, pick up two, pick up four, and wild.

Rules of Play

1. To begin play: deal out seven cards to each player, and flip one card over from the leftover cards for a discard pile. The remaining cards should be placed facedown for a draw pile. The person sitting to the left of the dealer plays first.
2. You can play any card from the same category: to play means to place down a card that is similar to the card already in the discard pile.
3. Star cards change the category in play.
4. Skip cards skip the person next to the holder.

https://doi.org/10.7330/9781646421947.c012

5. Reverse cards reverse the order of play.
6. Pickup cards require that the next player pick up the number of cards designated.
7. Wild cards change the category in play.
8. If you can't put down a card, you have to pick up another.
9. If you get down to your last card, you have to say "tutor" before the next play; failure to say "tutor" requires that a player draw another four cards.
10. Players have to read their card as they put it down. This ensures that the card matches the category in play.
11. Once a player is out of cards, the hand is over.

Winning/Completing the Game

To win, a player must discard all cards before any other player. Players can be ranked by who runs out of cards, or they can choose to score the cards left in each hand. Category cards are worth 5 points, non-wild word cards are worth 10, and wild cards are worth 20.

Variations

For longer games, double the suits, use more training terms and skills, or combine different decks. Other variations of the rule cards can also be created and added. This game can also teach grammatical tenses, sentence and word parts, or anything else that can be grouped into categories.

Rationale for the Game

Without active listening, whether audible or visible, authentic communication cannot occur and the tutor risks subordinating the writer's concerns in a session. Isabelle Thompson and Jo Mackiewicz (2014) assert that "learning typically occurs within a conversational context, and along with stimulating understanding, questions are vital linguistic components of an educational conversation" (39). Active listening strategies provide tutors with a model for questioning and paying attention, encouraging dialogue, and negotiating meaning as the conversation unfolds.

Active Listening Uno may not teach students to become perfect listeners, but it creates a concrete means of identifying the skills necessary to listen better. Regardless of whether a tutor can be taught to listen, as Tracy Santa (2016) equivocates, "Active listening seems central to establishing an ethos of cooperation and shared responsibility in writing center tutorials" (8). Therefore, tutor training necessarily centers on understanding active listening in a variety of modes, both verbal and nonverbal, audible and visual.

Students remembered the experience and the foundational concepts they used to create connections during the game because of both the active nature of game play and the swift pacing. Because the game requires players to react quickly and remain attentive, it also helps conceptualize pacing strategies transferable to tutoring sessions.

As Carl Glover (2006) asserts, tutors can learn by observing "body language [which] might signal some form of intellectual breakthrough" (17). Likewise, when changing suits or attempting to read and predict other players' moves or hands, tutors potentially learn to anticipate and identify these moments more clearly.

Regardless of whether students are actively listening, either to verbal or nonverbal cues, while playing or simply reviewing concepts for asking questions in a session, *Active Listening Uno* energizes tutor training sessions, activates learning, and centers an important skill.

REFERENCES

Rules adapted from https://www.unorules.com/.

Glover, Carl. 2006. "Kairos and the Writing Center: Modern Perspectives on an Ancient Idea." In *The Writing Center Director's Resource Book*, edited by Christina Murphy and Byron L. Stay, 13–21. New York: Routledge.

Santa, Tracy. 2016. "Listening in/to the Writing Center: Backchannel and Gaze." *Writing Lab Newsletter* 40, no. 9–10 (May-June): 2–9.

Thompson, Isabelle, and Jo Mackiewicz. 2014. "Questioning in Writing Center Conferences." *Writing Center Journal* 33, no. 2: 37–70.

13

HEADS UP!
Asking Questions and Building Vocabulary

Stacey Hoffer

Objectives

- To engage higher-order thinking skills.
- To add a sense of play in training.
- To help build skills for reading nonverbal cues.
- To help encourage lateral thinking.

Number of Players

Minimum of two people to play. Teams and large groups can also play.

Materials

- Index cards.
- Marker.
- 60-second timer.

Starting the Game/Setup

1. Using the index cards and marker, create a stack of playing cards. Each card should have one word/phrase on it. The words should be terms, theorists, types of writing styles, types of tutoring styles, or other key content from the tutor education course, readings, or concepts you've been discussing in your centers.
2. Place these items face down on a table.
3. Place a 60-second timer on the table.
4. The timer starts and the game begins when a player picks up a card and holds it up to their forehead to begin guessing.

Rules of Play

1. Divide the group into at least two teams of at least two people.
2. A player picks up a card from the pile and puts it on their forehead.
3. Other players on the team try to get the player to guess the word on their forehead before time runs out.

https://doi.org/10.7330/9781646421947.c013

4. Teammates can use any verbal or nonverbal strategy to encourage the player to guess; however, clues must not include the word(s) on the card.
5. Players cannot look at the card before selecting it.
6. Rounds cannot exceed 60 seconds.
7. Players can discard and select a new card after 30 seconds.
8. Incorrect and discarded cards must be shuffled into the stack after each round.

Winning/Completing the Game

Once all the cards have been used, players tally correct guesses. Whichever team has the most correct guesses wins.

Playing Tips

Try not to pass too often. Also, nonverbal clues are often more quickly understood.

Variations

This game can also be played in pairs using sticky notes. In this case, there isn't an established winner, but everyone has the opportunity to keep guessing until they figure out their term, style, or theorist.

Rationale for the Game

This game demonstrates the negotiation of meaning necessary for successful tutoring sessions. In their comprehensive study of questioning methods in tutorials, Isabelle Thompson and Jo Mackiewicz (2014) identified two types of questions experienced tutors ask: "(1) questions to establish common ground, including questions tutors asked to be sure they understood the assignment and the students' conference goals and to evaluate the students' understanding and (2) questions to lead and to provide scaffolds for students, aimed at moving students along in their brainstorming and revising" (45). This game requires that players ask scaffolded questions and think laterally about the concepts covered in training to determine their terms. Furthermore, this game allows for activation of critical thinking skills and provides an excellent foundation for further discussion of the concepts covered and the potential connections between them once the game is complete.

REFERENCES

Thompson, Isabelle, and Jo Mackiewicz. 2014. "Questioning in Writing Center Conferences." *Writing Center Journal* 33, no. 2: 37–70.

14

AND NOW PRESENTING
Marketing Writing Center Identities

Rachael Zeleny

Objectives

- To employ role play as a means of refining interpersonal skills with peers and clients.
- To enhance staff building.
- To improve tutor knowledge of writing center function.

Number of players

Six to sixteen players.

Materials

- Several boxes of "dress-up" material (e.g., capes, tiaras, hats, gloves, shields) in the front of the room.
- Blank templates for storyboards.
- Graphic organizer.
- Links to model writing center videos such as this Mythbusters promo video (Highland Center Community College 2010).
- Optional: a prize for the winner.

Starting the Game/Setup

1. Split players into, at minimum, two even teams.
2. Place the boxes of dress-up items in the front of the room.
3. Hand out storyboards, scripts, and graphic organizers.
4. Set up a computer projector for viewing models.

Rules of Play

1. Explain the concept of pastiche (mimicking the style and conventions of a certain genre) and show the Mythbusters writing center video (Highland Center Community College 2010).
2. Ask each group to complete a graphic organizer to determine key pieces of information they want to highlight about your specific writing center.

https://doi.org/10.7330/9781646421947.c014

3. Each team should brainstorm possible motifs for their video (e.g., *The Bachelorette, Harry Potter, The Avengers*).
4. Ask one representative from each group to come to the front of the room.
5. Set a timer for one minute, and ask the chosen representative to choose four props/accessories from the dress-up boxes that will enhance the group's chosen theme.
6. The representative will take these items back to the group so they can consider how to incorporate props into their script using the storyboard template.
7. Using the sample elevator pitch as a model (see supplemental materials), the group should prepare a pitch to present to the rest of the groups.

Winning/Completing the Game

Each group will take turns acting out/presenting their pitch. The winner will be chosen by either asking students to vote or asking external members to vote, such as faculty in other departments.

Playing Tips

Try to make at least four boxes of dress-up materials, and create a variety of items for each box. Avoid a box that is all "princess items" or any similar theme. Try to explain the concept of pastiche prior to the workshop, and perhaps have a discussion about what pop culture motifs have longevity as opposed to motifs that are relevant only for a moment.

Variations

The game could be varied by creating marketing materials such as a poster that would mimic a movie poster instead. Potential follow-up: students will produce a film to be featured on the writing center website.

Rationale for the Game

And Now Presenting is a role-play, media production activity that functions not only as a team-building activity but also as an opportunity to shape the writing center identity. The origin of this game was twofold: first, members of our writing center wanted a fun way to reduce anxiety surrounding writing center visits while also giving tutors the opportunity to become better advocates for the type of learning that occurs in the writing center. As demonstrated by the efficacy of writing fellows programs (where students are embedded in certain classes to collaborate with faculty to support the student writing process), student-led advocacy is the most effective means for "turning the biggest skeptics into the biggest advocates" (Furlong and Crawford 1999, 26). We wanted a way of introducing the center and the tutors to our students, to advertise services, and to correct misconceptions about the writing center in a playful way.

By asking the tutors to create the narrative by using a pop culture pastiche, we were asking them to think about the students on several levels. As Kate Warrington (2011) observes, "Writing centers commonly say that they are 'open to all students,' but not all students are equally likely to visit the writing center" (151). Therefore, tutors are required to consider who is unlikely to visit the writing center and why. They consider which pop culture references might appeal to our student body and how they make the writing center more accessible. They have to decide what would be compelling to a student as a reason to come into the writing center.

Ultimately, it is not especially important if the marketing video is ever made. First and foremost, the activity is fun. Tutors are often well accustomed to reality television shows such as *Nailed It* or *LEGO Masters* in which a group of individuals are asked to produce something in a compressed time period. Asking them to do so with these video storyboard pitches creates instant bonding. Second, the pop culture allusions become yet another resource in their toolbox when talking with students about why the writing center is an accessible place. For instance, I once heard a tutor talking to a student in my class and saying that if they did not find a tutor they "clicked with" during the first visit to just "not give that one a rose" (a reference to ABC's *The Bachelor*) and try someone else.

Furthermore, this process creates empathy for the writing center client because many tutors choose to role play the reasons students are unlikely to attend as part of their marketing video. As an example, one group used the various characters from *Harry Potter* to emphasize how each personality would need very different things from each tutor, ranging from Hermione wanting a spell to fix her grammar to Ron exhibiting anxiety about how to start his paper and throwing up slugs. Finally, it reinforces the key functions of the writing center into soundbites tutors can use to de-stigmatize the center, which is crucial since if students "think that going to the writing center is stigmatized, then they will choose not to visit, even if they genuinely want help with their writing" (Salem 2016, 147). Thus this process is less about producing a video than it is about creating engaging, empathetic, and well-spoken ambassadors for the center.

REFERENCES

Furlong, Katherine, and Andrew B. Crawford. 1999. "Marketing Your Services through Your Students." *Computers in Libraries* 19, no. 8: 22–24.

Highland Center Community College. 2010. "Mythbusters." https://www.youtube.com/watch?v=5GMKqmYL8YM.

Salem, Lori. 2016. "Decisions . . . Decisions: Who Chooses to Use the Writing Center?" *Writing Center Journal* 35 (2): 147–71. http://www.jstor.org/stable/43824060.

Warrington, Kate. 2011. "Bolstering Writerly Instincts: Using Role-Play to Help Tutors Address Later-Order Concerns." *Praxis: A Writing Center Journal* 8, no. 2. doi: 10.15781/T2TG70.

SUPPLEMENTAL MATERIALS

WRITING CENTER PROMO: ELEVATOR PITCH

1. Give an overview of your **center's function and goals**.
2. Give an overview of **hours**, **staff** numbers, and methods for making an **appointment**.
3. Address the **goal of your video**.
4. Explain what **pop culture reference** you are making and how this narrative would help emphasize a particular facet of the writing center's function.

Sample Pitch

Our video will attempt to address the various reasons students provide for not going to the writing center. Using Disney's *Inside Out,* we will use the various characters, Anxiety, Logic, Procrastination, and Pride, to personify and debunk the various myths students create for why they do not visit the writing center.

Our video will begin with a closeup of a student's head, then pan the camera closer, then show a screen of text that says "Thinking about Writing Centers . . . Inside Out!" Then pan to screen of four characters dressed in different colors with these words pinned to their shirts: logic, procrastination, anxiety, and pride.

ANXIETY: I can't write this paper, I'll never write this paper, even if I write this paper, it will be terrible!!

LOGIC (in robot voice): The University of Delaware's writing center offers free one-to-one tutoring, providing all UD students with the opportunity to work on and improve their writing.

ANXIETY: But I have no idea what happens there. What if they make me take a quiz on grammar? What if they're judging me?

LOGIC: Tutorials are conversation-based. You can expect to read through your writing with an experienced writing tutor, talk about what is working well, and discuss what could improve. The center offers assistance with all types of writing (academic or non-academic) and works with students at any stage of the writing process. The tutors do not edit papers or simply check for grammar errors.

PROCRASTINATION: But I don't have time. Besides, I don't even know how to make an appointment.

LOGIC: Appointments start on the hour and last approximately fifty minutes—limited to two appointments a week. Please make an appointment using the online scheduler. The website has more details.

PROCRASTINATION (staring at phone screen): But I don't have anything written yet, so it's a waste of time for me to make an appointment. Besides, I'm sure there is a kitten somewhere doing something cute on a video . . .

LOGIC: You can go to the writing center at any stage of the writing process. You can even go just to look at the prompt and make an outline. Please proceed to ignore procrastination.

PRIDE: So if I go to the writing center now, maybe I will actually like this paper I'm writing . . . ?

LOGIC (rolling eyes): Yes, that's what I've been saying for years now.

PRIDE (sitting in front of computer to make appointment): Lalalala. This will be the best paper ever. I'll probably win a prize. Maybe they will publish my paper immediately . . .

LOGIC: Let's just aim for a good grade, okay?

Close-up on stapler that is opening and closing like a mouth: "The End."

Brainstorm/Notes

WRITING CENTER MARKETING VIDEO

Graphic Organizer

Key Information to Include

Services offered

Location/hours of operation

How to make an appointment

What to expect from an appointment

Who the tutors are (what makes them qualified)

How to prepare for an appointment

Movie Title:____________________________________

Information	*Symbolism*
EX: How welcoming the tutors are versus people who make fun of those who go to the center	EX: Janis and Damien accepting Cady as their friend (*Mean Girls*) versus the Plastics that make her an outcast

15

ESCAPE THE SPACE

Building Better Communication with Peers through Problem-Solving Situations

Christina Mastroeni, Malcolm Evans, and Richonda Fegins

Disclaimer

We would like to inform our audience that this instruction set is specifically designed for the Game Master (head organizer). Spoiler alerts happen consistently for other players throughout this set.

Objectives

- To familiarize students and staff with the physical writing center environment.
- To cultivate teamwork among players.
- To facilitate good communication practices.
- To enhance critical thinking and problem-solving skills.

Number of players

Two to ten players.

Materials

Before you begin this instruction set, you will need the following items:

- A timer (e.g., phone, tablet, computer).
- A four-digit combination lock.
- A five-digit combination lock.
- A five-letter combination lock.
- Any four books with different titles and authors.
- Three boxes that can be locked with the three combination locks.*
- Pens, pencils, or both.
- Pad/paper (for note taking).

**Can substitute any item that has a similar locking structure (e.g., luggage).*

https://doi.org/10.7330/9781646421947.c015

Starting the Game/Setup

The Game Master(s) is responsible for setting up the game.

Setup Process

- Place the timer in a visible area.
- Clearly identify the space(s) that is(are) part of the game.
- Place the reference list (see Game Image 2) in an obvious place for participants to find when the game begins.
- Cut out each number from Game Image 3 individually (4, 7, 3, and 8). Assign each of the books a number, starting with the first source that appears on the reference list; tape the first number (4) inside the front cover of that book. Moving to the second source that appears on the reference list, tape the second number (7) inside the front cover of that book. Make sure each book on the reference list has a number included inside its front cover.
- Place the limerick (Game Image 4) in a box, locking the box with the four-digit combination lock and the combination 4738.
- Place the cipher (Game Image 5) in a second box, locking the box with the five-letter combination lock and the combination WRITE. If needed, hints are available (Game Image 6).
- Place the key image (Game Image 7) in the third and final box with the five-digit combination lock and the combination 37425.
- Hide the boxes around your space.
- 2–10 players begin together as they listen to the Game Master reading the introductory letter.
- The timer is set to twenty-five minutes and begins immediately after the introductory letter is read.

Rules of Play (Game Master)

How to Play

- Step 1: Read the letter containing context for the game's objective (see Game Image 1). Note: Review the goal of the game as a team so each player is communicative from the beginning. However, the players should not read the letter.
- Step 2: After the letter has been read aloud, start the timer. Note: Give you and your team a reasonable amount of time to achieve your goal. Be sure to discuss and agree on your expectations moving forward.
- Step 3: Give the participants hints as needed. Note: As the leader/Game Master, you need to be in tune with your team. When they begin to show signs of struggle (verbal or nonverbal), provide assistance. Here, you're leading by your position and by example.
- Step 4: Once the participants find the key image, they have successfully completed the game. Note: Achieving this step provides an assessment tool for you and your group. You'll be able to determine if you completed the task successfully by your group's predetermined standards.

Winning/Completing the Game

The participants solve the final clue and find the key image before time expires.

Playing Tips

Tip 1: Encourage participants to stay in the clearly identified game space.
Tip 2: Negotiate with players how involved they want the Game Master to be in providing hints.
Tip 3: Lessen the amount of time participants have to complete the game for an added challenge.

Rationale for the Game

Writing centers are creative, liberating, and collaborative spaces that foster and encourage interaction. Writing center associations such as the Mid-Atlantic Writing Centers Association and the International Writing Centers Association call for creative ways to encourage center staff to participate in a culture that is conducive to the cultivation of such an experience. Escape rooms, much like centers, encourage exploration and collaboration through questioning (Nicholson 2015). This approach to discovery and teamwork is reflective of the Socratic methods employed in tutor sessions. Thus using escape rooms for tutor instruction offers a unique situation in which players with equal authority work through the process of identifying their strengths and improving their weaknesses. The product of this game is taking these improvements and utilizing them in tutoring sessions to create an effective experience for both the tutor and the tutee (Corbett 2013).

This activity involves team building within the writing center environment. Adapted from popular escape room games, *Escape the Space* allows writing center staff to work together to familiarize the staff with the physical writing center space and resources, encourage staff to practice collaboration in an effort to achieve goals, promote interactivity and creative problem solving, reinforce training content used during sessions (i.e., citation styles), allow staff to become comfortable communicating in preparation for their center work, and encourage play in the writing center.

The re-creation of the game can be made using a tabletop version. This version will allow groups of participants to interact with the display in turn as they familiarize themselves with the concept of the game and how to incorporate it in their centers. Peer tutors and administrators will have the opportunity to play in our themed demonstration. We will provide puzzles and brain teasers that will challenge participants to collaborate in an effort to reach resolution while reinforcing the tools/strategies we use in our centers daily. We will also provide suggestions and themes for participants to take back to their centers.

REFERENCES

Corbett, Steven J. 2013. "Negotiating Pedagogical Authority: The Rhetoric of Writing Center Tutoring Styles and Methods." *Rhetoric Review* 32, no. 1: 81–98. doi:10.1080/07350198.2013.739497.

Nicholson, Scott. 2015. "Peeking Behind the Locked Door: A Survey of Escape Room Facilities." White Paper. http://scottnicholson.com/pubs/erfacwhite.pdf.

Vorhees, Dayton. 1902. "Limerick." *Princeton Tiger*, November, 59.

SUPPLEMENTAL MATERIALS

Letter: To be read by Game Master only (Game Image 1)

Puzzle 1:

- Reference List Template (Game Image 2)
- Four-Digit Combination (Game Image 3)

Puzzle 2:

- Limerick (Game Image 4)

Puzzle 3:

- Cipher (Game Image 5)
- Hints to Solve Cipher (Game Image 6)

Puzzle 4:

- Picture of a Key (Game Image 7)

Game Image 1: Letter

The Game Master should read this before beginning the game. The players should not read this letter.

Our heroes of the writing center have been trapped by their archnemesis, the Pun Master. Pun Master has yet to defeat our heroes, despite his many attempts. However, he is still confident that he will be able to finally defeat the team and solidify his status as a genius criminal mastermind. He has infiltrated the writing center and set up a trap for our heroes. To escape, our heroes will have to use all their powers of deduction and reasoning to unlock/uncover all the locks/puzzles needed to defeat Pun Master. If the team fails to escape, they will be crushed under the weight of 40 million wrinkled sheets of paper taken from plagiarized student papers. Oh the Iron-y!

The team has only twenty-five minutes until the alarm sounds and the papers are released. In addition to the puzzles/locks, Pun Master has also locked two team members together (knowing that dastardly villain, he has most likely hidden a key somewhere in the vault). Another team member is not able to speak (but can write), while another has been blasted by his rhyme ray and can only speak in rhymes. Will our heroes escape in time? Do they have the fortitude and problem-solving skills they need? Good luck, team!

Game Image 2: Reference List Template

To use, insert the information from your chosen references in the correct APA format (as shown on the template in Game Image 2). Add three more books in addition to the four you have already chosen; the additional three will help throw the participants off their game.

Escape the Cube 22

References

Author, A. A. (Year of publication). Title of work: Capital letter also for subtitle. Location: Publisher.

Author, A. A. (Year of publication). Title of work: Capital letter also for subtitle. Location: Publisher.

Author, A. A. (Year of publication). Title of work: Capital letter also for subtitle. Location: Publisher.

Author, A. A. (Year of publication). Title of work: Capital letter also for subtitle. Location: Publisher.

Author, A. A. (Year of publication). Title of work: Capital letter also for subtitle. Location: Publisher.

Author, A. A. (Year of publication). Title of work: Capital letter also for subtitle. Location: Publisher.

Game Image 3: Four-Digit Combination

4	7	3	8

Game Image 4: Limerick (Vorhees 1902)

Place this limerick in a box. Lock the box using the four-digit combination (from Game Image 3): 4798. The answer to the limerick (seen in the italicized and bolded letters in Game Image 4) is WRITE.

"There once was a man from Nantucket
Who kept all his cash in a bucket. But his daughte***r***, named Nan,
Ran away w***i***th a man,
And as for the bucke***t***, Nantuck***e***t."

Game Image 5: Cipher

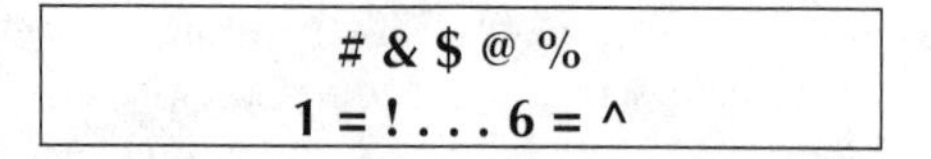

Place this cipher in another box. Lock the box using the five-letter combination lock, using the combination WRITE. The answer to this cipher is 37425.

Game Image 6: Hints to Solve Cipher

)	0	%	5
(	9	$	4
*	8	#	3
&	7	@	2
^	6	!	1

Game Image 7: Picture of a Key

Find a picture of a key online and print it out, then put it inside the final box. Lock the box using the five-digit combination lock, using the combination 37425.

16
LEVEL UP

Alyssa Noch

Objectives

- To get the writer(s) past whatever part of the writing process (e.g., brainstorming, editing, revising) they identify as being stuck on.
- To get the writer(s) and the facilitator(s) to team up and aid the advancement of the writer(s) to the next stage, or level, of the writing process—to level up.
- To assist in a better understanding of the writing process and one's own writing skills using metaphors.
- To give the writer(s), through game language and play, the means to articulate and think about their skills in a different way.
- To, when more than one writer plays this game, facilitate effective group work by identifying individual group members' skills and developing their relationships with each other.

Number of Players

Two to five players who must be grouped into two categories:

- o Writer (driving/main character). This is the student, consultee, or other individual who is seeking aid on a writing project. No more than three writers should play at a time, and they must all be working on the same writing project.
- o Facilitator (support/side character). This is the consultant, faculty, staff, tutor, or other authority who takes on a supportive role in helping the writer. It is best that there are no more than two facilitators while playing. Facilitators are responsible for leading the other players in game play, promoting positive engagement and creativity.

A game must include at least one player from each player category to be played. Therefore, this game must be played with a minimum of one writer and one facilitator.

Materials

- Print out images of characters and items from video games, books, movies, or other media, or print out/create your own characters and items. Each writing center, tutoring center, or other organization can

https://doi.org/10.7330/9781646421947.c016

generate their own characters and items. They can add to their collection of characters and items as time goes by, based on their writers' needs.

- Characters: Avatars the players can choose to play as during the game. Examples are:
  - Characters from video games such as *Mario Kart* and *Super Smash Bros.*, such as Mario, Luigi, Princess Peach, Sonic the Hedgehog, Link, Zelda, and others.
    - Characters from book series (such as *Harry Potter*), famous authors (such as William Shakespeare), or historical figures (such as Harriet Tubman).
  - Flashy Fox, a made-up character that is known for their flashy vocabulary.
  - Brainstorming Bard, a made-up character that is known for their brainstorming prowess; their skill may also be displayed through an item in their possession.
  - Unicorn of the Lost, a made-up mythical creature known for leading writer(s) out of the darkness of tricky writing levels.

  Items: Representations of certain abilities the writer(s) and the facilitator(s) have or can acquire to further cement the role-playing aspect of this game. Here are a few examples:
    - A fireball (an idea participants can take from a video game such as *Super Mario Bros.*) that can be used on a brainstorming level meant for the rapid-fire generation of ideas, getting all the writer's thoughts down on a whiteboard or a piece of paper.
    - A blue shell with wings (an item from *Mario Kart*) that could be used in the revision process to signify a cutting or reorganizing of whole paragraphs in an essay.
    - A wizard hat (an item that belongs to no video game in particular) that can be used during the editing process, symbolizing the astuteness required of someone editing a piece of writing when looking for grammar mistakes or other technical errors.
    - A sword (a generic item that belongs to no video game in particular) that can be used during the revision process to symbolize the slashing, or cutting out, of unnecessary words, sentences, and paragraphs in a piece of writing.
    - Headphones (an everyday item imbued with a new meaning) that can be used to symbolize the use of meter, rhythm, or rhyme in a piece of poetry or music.

Table or other areas you can set these characters and items on, or a whiteboard, portable whiteboard, paper, or chalkboard setup where you can draw characters, items, and levels (may also include dry-erase markers, chalk, eraser, pens, pencils, or damp cloth).

Starting the Game/Setup

1. Identify Level

   The players of this game must identify where the writer, or writers, need(s) aid in their writing process (e.g., brainstorming, revising, editing). The writer(s) in need of aid must first be asked by the facilitator(s) where they need help with their writing. Finding out what stage of writing a person is stuck on determines the level or next stage of writing they need to level up to. The writer(s) determines these levels. It is up to the facilitator(s) to help the writer(s) identify the levels and aid them in leveling up to their desired level.

2. Pick Characters

   The writer(s) must pick their character(s) to play as. These characters can be from video games, books, movies, history, or one's own imagination. The chosen character(s) by the writer(s) are then to be perceived as the main character(s) in this game. The facilitator(s) playing this game must choose their character(s) secondary to the writer(s). All facilitator characters are to be perceived as supporting or side characters in the game and should be chosen with that in mind. This means their goal is to help guide the main character(s) through the stage of writing (or the level) they are on; their character(s) should reflect their supporting nature with abilities they inherently possess and through the items (e.g., symbols of different writing skills) they carry. These side characters can be added to or subtracted from the game as different items, or skills, are needed. Side characters are only allowed to be equipped with a few items at a time that may be necessary for the main character(s) to achieve the goal of leveling up to their next designated writing level.

3. Identify Items

   Items may be gifted to the main character(s) by the side character(s) or "used" directly on the piece(s) of writing belonging to the main character(s). The main character(s) can also start with or acquire some of their own items, depending on how proficient they feel they are at a certain part of the writing process or by gaining some of the items needed for the level through the course of the game. The facilitator(s) can help the writer(s) identify what their character's items should be at the start of the game or award these items to the writer(s) during game play as they feel these items, or skills, are acquired.

4. Game Start

   Once all initial characters and items are assigned and the level(s) identified, the game can begin.

Rules of Play

1. There must be at least two players in each game, one identified as a writer and one identified as a facilitator.

2. Identify the level of writing the writer(s) needs help with (e.g., brainstorming, revising, editing).
3. Assign the game characters and items to the participating writer(s) and facilitator(s). No more than two items should be given to the writer(s) seeking consultation; these items should reflect areas of the writing process or other writing skills the writer(s) identifies as their strongest. The facilitator(s) should have no more than three items at the start of this game. These items can reflect either their best writing skills or the weakest identified skills of the writer(s). If more than one facilitator and writer are playing the game, items should try to be as diverse as possible and not overlap with each other.
4. Work through the writing level. This is done by working through the piece(s) of writing presented by the main character(s). For example, the writer identifies that they need help on the revision stage of their essay. The facilitator, as the side character, helps the writer work through this level by reading their piece aloud, as is often done in regular writing consultations.
5. When a writer's struggle is identified during game play (this struggle could be, for example, repetitive word usage or wordy sentences), one of the participants' items should be introduced or used as a way to display and explain the skill needed to move on to the next writing level. These items can be transferred from the side character(s) to the main character(s) if an understanding and improvement of the skills of the main character(s) is felt to have been archived (i.e., mastery of the item has been taught by one player to the other).
6. More than one side character can be used at a time, and they can even sub in and out with one another if needed for items the other facilitator(s) may have.

Winning/Completing the Game

The game can be completed in one of two ways:

1. When the writer(s) and facilitator(s) have met the goal the writer(s) selected. This means the writer(s) feels they have leveled up to the next stage of writing.
2. When the writer(s), facilitator(s), or both have decided they need new characters to change up the game or need a break in game play. The context of the situation will help establish a time to stop game play. For example, if the game is being played in a writing center where there is an established time limit for consultations, then the game play end is determined by this time limit.

The prize for winning is a further understanding of the writing process as well as a better grasp of the strength of one's colleagues and oneself in the field of writing. Plus, if the center allows writers to keep the items they gained in the process of leveling up, the writers can leave with a tangible reminder of their game play.

Playing Tips

1. Don't stress too much about not knowing all the characters and items used for this form of role play from specific video games, books, movies, and other media. They are not consequential to the effectiveness of game play as they are just a useful framework to use so players are not starting from scratch when making up their own game characters and items.
2. Facilitators at a writing center, tutoring center, or other venue should feel free to create their own game characters and items along with the abilities those items are meant to display in furthering the advancement of the writing process. Their repertoire of characters and items can be added to and changed over time based on the needs of each center and with regard to their clientele.

Variations

This is a general outline for game play, but as *Level Up* is played, it can be subject to change and advancement in different ways. This is because *Level Up* is meant to be a flexible and self-determined game by the writer(s), with consideration given to suggestions and input from the facilitator(s). Each writing center, tutoring center, or other venue can always set limits to this or create their own version of *Level Up* for their centers.

This game is not a formal variation of any other writing game out there, although variations can be made with it through the characters and items chosen to be used for this form of role playing.

Rationale for the Game

The slang term *leveling up* is used as a metaphor in the context of writing centers, tutoring centers, and other venues to refer to the advancement of the writer(s) to the next stage, or "level," of the writing process. Used in the game *Level Up*, it equates the idea of "leveling up" in writing centers, tutoring centers, and other venues with "leveling up" in video games (and other games with noted levels to them such as *Dungeons & Dragons*) as players advance on to higher levels after completing a set number of tasks. In writing centers, these tasks that must be completed to level up represent a stage of the writing process. By completing a level, or stage, of the writing process, the writer(s) will have moved a level up or moved on to their next stage in the writing process.

As a game, *Level Up* shows writers that they do not have to write alone and that they have skills that can be used to write. Reaching out to a tutor for help can be better than trying to look up writing tips online, as these tips can vary and often be confusing and stressful for the novice writer to try to understand. To avoid this confusion and stress for the writer(s), as Muriel Harris (1995) says, tutors are needed as they "are there to help reduce the stress, to overcome the hurdles set up by others, and to know more about writing than a roommate or friend, maybe even as much as their [the writers'] teachers" (29). In accordance with this, *Level Up* encourages writers to seek human interaction and to lean on facilitators who can help them accomplish their goals of leveling up to the next stage of their writing process.

This game strives to be student-centered. This hearkens back to one of the oldest maxims in the field concerning writing centers, written by Stephen North (1984): "Our job is to produce better writers, not better writing" (438). For this reason, this game aids those with varying learning styles and preferences who might find learning and writing through metaphors and games effective.

For more than one struggling writer working on a group project, this game identifies and highlights each individual writer's skills. Every skill a writer is identified as having is valued in the game through its depiction as an item. For example, the skill of brainstorming can be represented as a fireball or a thunderstorm in the game. These skills, coming from each character, are designed to be utilized, shared, and communicated to the rest of the group through the course of the game. Jennifer M. Hewerdine (2017) notes that "collaboration, when it incorporates diverse people, provides a means of sharing and understanding diverse knowledge, thinking, and experiences" (35). It is my hope that *Level Up* will draw attention to this value of diversity and teamwork—fostering respect, the furthering of communication, and the sharing of work within a group through its game play, creating a safe and productive space for all those involved.

REFERENCES

Harris, Muriel. 1995. "Talking in the Middle: Why Writers Need Writing Tutors." *College English* 57, no. 1: 27–42.

Hewerdine, Jennifer M. 2017. "Conversations on Collaboration: Graduate Students as Writing Program Administrators in the Writing Center." PhD dissertation, Southern Illinois University, Carbondale.

North, Stephen. 1984. "The Idea of a Writing Center." *College English* 46, no. 5: 433–46.

17

WRITING AND ROLE PLAYING

Mitchell Mulroy

Objectives

- To create a writing style definition that characterizes a writer's strengths and challenges.
- To apply this definition to a writer's assignment or project in a way that caters to that particular writer's identified strengths and challenges.
- To help writers personalize their approach to writing.

Number of Players

At least two players.

Materials

- A piece of paper.
- A writing utensil.

Starting the Game/Setup

The consultant should begin the game by identifying the writer's definitive class. In tabletop role-playing games, a "class" is a chosen group of character traits that make up a kind of archetype. For instance, someone with high intelligence who casts spells would be a wizard, while a high-strength character would be a fighter. Choosing a particular writer's class is done by breaking down elements or preferences of the writer's style into different categories. Perhaps someone who writes and outlines and prepares for an essay before actually writing it is a researcher and someone who starts writing the assignment without an outline or a plan is a director. Maybe one writer describes imagery with a detailed, passionate bent, in which case they would be a writer of the "artistic" or "romantic" class.

Rules of Play

1. Identify the writer's unique style and focus.
2. Use this information to create a unique class for the writer.

https://doi.org/10.7330/9781646421947.c017

Winning/Completing the Game

The writer wins when the consultant and writer create a class for the writer while also identifying ways to play to those strengths and weaknesses inherent in their designated class when completing writing projects.

Rationale for the Game

Nancy Chick (2019) examines the popularity and criticisms of learning styles and how they are used by teachers. She specifically connects learning styles themselves to a concept of meta-cognition, citing its wide recommendation for use in classrooms. Meta-cognition also applies to a concept of a writing class. Writers will have to directly consider their own process and what differentiates it from others, similar to the way people look at and identify their different learning styles. For instance, writers can define their writing style through meta-cognition to benefit in the same way learning styles benefit students in classrooms. Using classes as a baseline, the consultant is able to give the writer suggestions that more accurately match their perceived style. This might have the added effect of making insecure writers more accepting of criticism when they realize how it fits into the larger aspect of their writing.

By defining their own writing style in relation to others in a definable form, writers will have a greater understanding of their strengths and weaknesses. It may also allow writers to take pride in their unique skill sets. Many writers come to writing centers unsure of their own ability and disparaging of their own writing process. This is similar to the way many first-time role players can feel isolated and nervous participating in their first role-playing game. However, strongly identifying with one or more sets of traits their imagined role-playing character has often results in the individual becoming more comfortable with acting out their role in the game. The difference here is that the classes we create in this scenario would represent an individual's own interests and skills as a writer rather than the fictional traits of other characters they want to represent, with the goal of making them more comfortable with their own writing style and aware of how to write most effectively.

REFERENCES

Chick, Nancy. 2019. "Learning Styles." *Vanderbilt University*, November 6. cft.vanderbilt .edu/guides-sub-pages/learning-styles-preferences/.

SUPPLEMENTAL MATERIALS

Sample Classes

RESEARCHER: The act of writing is a culmination of different types of preparation. Staying organized while creating forms of pre-writing can drastically help the creation process.

DIRECTOR: You need your thoughts on the page as soon as you get them. Figuring out what you're writing as you're writing it is an engaging exercise that ultimately leads you to great conclusions.

ROMANTIC: Romantics are likely to create long descriptions of imagery and emotion. They are generally comfortable with creative writing while feeling limited by the formalities of professional writing. The descriptive nature of a romantic means they often "over-explain" in essay writing but are also able to form elegant sentence structures.

REALIST: Realists tend to think quickly and rationally about problems using input from somewhere else, and as a result they might feel that creative writing doesn't provide enough information from which to frame their writing. The analytical nature of a realist means they often go through vital information quickly without explaining their points but are also able to explain their points concisely.

18

WRITING ON THE WALL

Elysse T. Meredith and Miriam E. Laufer

Objectives

- To engage students, faculty, and staff in writing through play.
- To create a physical space that acts as a locus for community interaction.
- To communicate and develop relationships in the center both synchronously and asynchronously.

Number of Players

This game is suitable for any size group composed of staff, students, faculty, tutors (peer, professional, and faculty), and student assistants. The facilitators must take a lead in playing and encouraging engagement.

Materials

- A wall or section of a wall painted with whiteboard paint or a portable whiteboard, depending on space and funding.
- Black dry-erase markers (colored dry-erase markers do not erase well from whiteboard paint).
- Eraser or damp cloth.

Starting the Game/Setup

If a wall has been painted with whiteboard paint, place black dry-erase markers visibly nearby. If using a portable whiteboard, place in a visible area with black dry-erase markers. Verbally introduce the game to participants as needed, or post an explanatory sign.

Rules of Play

1. Each week, the facilitating staff member writes a prompt on the whiteboard (see supplements at the end of the chapter). This begins a round.
2. Throughout the week, staff, students, and faculty in the writing center respond. All players are expected to add anonymous comments.

https://doi.org/10.7330/9781646421947.c018

3. Throughout the week, participants may also urge others to contribute, try to guess comments' authorship, and reply to others' comments.
4. At the end of the week, staff take pictures of the board to share on social media, in newsletters, or otherwise to continue engaging in conversations. This ends the week's round.
5. After pictures are taken, erase the board and begin a new round with a new prompt.

Winning/Completing the Game

As a free-form game, there is more than one way to complete the game; indeed, each player can decide how they win. Here are some ways of winning:

- Write the most responses.
- Write the most/fewest appropriate responses.
- Write the most-discussed post.
- Write the most–replied-to post.
- Write a post that no one can guess the author of.

The real prize is to stimulate shared conversation and communication among all the moving parts of the center to build a stronger team.

Playing Tips

- Encourage full-time staff and regular tutors to write several comments anonymously and in different handwriting, which will embolden other players to write.
- Ask players to suggest prompts once the game is established.
- Choose prompts that jumpstart or continue desirable conversations.
- Use recommended readings and on-campus events as inspiration for prompts.
- After writing your own response, reply to another one.

Rationale for the Game

Writing on the Wall is a free-form game that serves as a locus for team building, professional development, and student engagement. The genesis for the wall was our desire to cultivate a trickster mind-set about writing in our center, as with the *Scrabble*-game-during-tutoring that is described in the chapter "Trickster at Your Table" in *The Everyday Writing Center* (Geller et al. 2007, 17). We wanted a similar physical demonstration that writing can be fun, informal, and transgressive. Furthermore, the writing wall also functioned as a manifestation of a welcoming writing center space. As Jackie Grutsch McKinney (2013) confirms in *Peripheral Visions for Writing Centers*, "The idea that a writing center is—and should be—a cozy . . . place is . . . firmly entrenched" (20). Although McKinney rightly problematizes this concept and its concurrent assumptions regarding student preferences, we believed the board would adapt to students' interpretations and interests while representing what we are about: writing.

However, trickster-like, when students resisted participating, the game evolved into a locus for staff development instead. This game is particularly useful when writing center employees, such as ours, fall into multiple different and often overlapping categories (full-time, part-time, administrative, professional, faculty, peer, student) that require complicated schedules. The whiteboard wall can help build rapport despite incongruous schedules, supporting staff development along with community outreach. If prompts are used to jumpstart or continue conversations, the wall can enhance professional development. For example, we followed recommended reading on linguistic diversity with a prompt about languages and dialects. "Comment wars" can be conducted asynchronously, enhancing staff play, creativity, and team building; in our center, we have watched faculty and students wield whiteboard markers as they deftly jab at one another across time. In addition, the more employees that participate, the more likely students are to respond. As Anne Ellen Geller and her colleagues (2007) elucidate, the malleable nature of the game, which plays with writing and meaning in the center, helps staff and tutors nimbly adapt to the trickster-like unpredictabilities of tutoring pedagogy and practice.

Overall, *Writing on the Wall* builds a creative, playful atmosphere for students that also enhances the work environment. The wall also serves as a physical location that allows students, staff, and faculty to engage with one another asynchronously. In a busy center, this game connects all the moving parts.

REFERENCES

Geller, Anne Ellen, Michele Eodice, Frankie Condon, Meg Carroll, and Elizabeth H. Boquet. 2007. *The Everyday Writing Center: A Community of Practice.* Logan: Utah State University Press.

McKinney, Jackie Grutsch. 2013. *Peripheral Visions for Writing Centers.* Logan: Utah State University Press.

SUPPLEMENTAL MATERIALS

Sample Prompts

- What is your favorite book or series?
- In honor of Valentine's Day, who is your favorite fictional couple?
- If I had a million dollars, I would . . .
- What languages or dialects do you speak?
- Write a haiku.
- What is the most important grammar lesson you have learned?

- What makes you happy?
- If I had a time machine, I would . . .
- What are you watching on Netflix?
- Tell us about an act of kindness.
- What quotes or sayings give you courage?
- How would you describe this semester in one word?
- Share your advice for getting through final exams and papers.
- Tell us what prompt you would choose.

19

ONE-WORD PROVERBS

Katie Levin

Objectives

- To enact valued writing center practices of listening, collaboration, and play: because of the game's high speed and low-stakes nature, staff must overcome individual perfectionism—*let me pause while I ponder the best possible word I can add here*—and instead focus on facilitating a successful collaborative experience.
- To reveal the power of writing and sharing in a low-stakes setting.
- To create and embrace texts that may include many different Englishes.
- To reveal the multiple possibilities that can arise from conversations among collaborators, depending on the roles each person plays: even though the same five writers are the coauthors of each proverb, the outcomes are as unique as can be.

Number of Players

This game requires minimally one group of three people, but it works best in multiple groups of four or five people.

Materials

- One sheet of paper for each group member (or group members can use a copy of the instructions, if you decide to hand those out).
- One writing utensil for each group member.
- A surface to lean on when writing (a binder, a book, a table).

Starting the Game/Setup

1. **Provide some context for the game**. I usually tell folks the following (here and below, *sample script is in this font*):

   I learned this game at the International Writing Centers Association (IWCA) Summer Institute, which is basically writing center camp. When Sherri Winans (2006) introduced it, she reminded us that this game is in the tradition of improv comedy, where two principles reign:

   - Say "yes" to everything. (This is also sometimes described as "yes, and"-ing.)
   - Our job is to make the other person look good.

https://doi.org/10.7330/9781646421947.c019

2. Explain why we're playing an improv game like this in a staff meeting. Note: Consider using the list from the objectives above.
3. Provide an overview of how the game works. Note: Not everyone has the same idea of what "proverb" means, so consider giving examples of proverbs or soliciting them from the whole group. Typical examples from a US context: "A rolling stone gathers no moss"; "The grass is always greener on the other side of the fence"; "You can lead a horse to water, but you cannot make him drink." Also, consider using the following language in your overview: *This is where the principles of improv come back: when we receive a word, our job is to say "yes, and" to it by reading the word, then writing down a follow-up word that takes the sentence where it wants to go. And one way we can make the other person look good is by providing those small words that help keep a sentence going (articles, prepositions). We don't always need to be the one who contributes the hilarious noun or the evocative verb; instead, we can add the little word that sets up the next person to shine.*
4. Provide the specific "how-to" details of the game. Project or hand out the instructions, and either summarize them or invite multiple people to read them aloud.
5. **Get people into groups of four or five**.

Rules of Play

See Supplemental Materials for step-by-step instructions.

Winning/Completing the Game

Truly, everybody wins. We usually play for just one round (because the game typically feels done after everyone has shared with the full group and because one round, including introduction of the activity and reading and sharing at the end, takes about 20 minutes)—but of course it's expandable for as much time as the full group wants to spend.

Variations

This has been used as a daily writing warm-up in our eleven-day dissertation writing retreat—typically, it's how we begin the final day, since it helps celebrate the community we've built.

Rationale for the Game

In addition to what's described in the objectives above, this game has many of the benefits Scott Miller (2008) identifies in his well-researched article making the case for the importance of play in the writing center. In particular, when staff members play this language-based game, they are "getting better and better at playing with language, writing wildly, donning and doffing diverse ethos with abandon[, which] is an essential element of learning to become a writer, of learning to live in a world where being *different kinds of writers* is essential for success, happiness, and empowerment" (38, original emphasis). Not only are all student writing support staff members consultants, but we are also developing writers, so this language-based game enriches us in both those roles.

The idea that the game is based in the principles of improv offers an additional benefit: as Jonathan P. Rossing and Krista Hoffman-Longtin (2016) argue, the principles of improv can be applied productively in faculty development, research collaborations, and classroom teaching to create the conditions for collaboration and trust. Rossing and Hoffman-Longtin suggest that improv games "invit[e] participants to practice skills such as listening, trusting one another, solving problems creatively, and adapting to uncertainty. Through experience, practice, and reflection, participants reinforce the importance and value of improvisational ways of thinking in the applied context" ("Applying Applied Improv"). Because all staff members engage in writing center work beyond the consultation—answering questions at the front desk; solving problems caused by last-minute absences; listening to, communicating with, and supporting each other when a tricky, uncomfortable, or otherwise unexpected situation arises—playing an improv game also serves our goal of consultants developing as leaders in the center.

Note: Adapted from Sherri Winans (writing center coordinator at Whatcom Community College; IWCA Summer Institute 2006 co-leader) by Katie Levin (co-director at the Center for Writing at the University of Minnesota–Twin Cities; IWCA Summer Institute 2006 participant).

REFERENCES

Miller, Scott. 2008. "Then Everybody Jumped for Joy! (But Joy Didn't Like That, So She Left)." In *Creative Approaches to Writing Center Work*, edited by Kevin Dvorak and Shanti Bruce, 21–47. Cresskill, NJ: Hampton.

Rossing, Jonathan P., and Krista Hoffman-Longtin. 2016. "Improv(ing) the Academy: Applied Improvisation as a Strategy for Educational Development." *To Improve the Academy: A Journal of Educational Development* 35, no. 2: 303–25. doi-org.ezp2.lib.umn.edu/10.1002/tia2.20044.

Winans, Sherri. 2006. "Games People Play." Workshop conducted at the International Writing Centers Association Summer Institute, Palo Alto, CA. July 24.

SUPPLEMENTAL MATERIALS

"One-Word Proverbs," a list of instructions to hand out formatted for players.

One-Word Proverbs

Materials

- One sheet of paper for each group member.
- One writing utensil for each group member.
- A surface to lean on when writing (a binder, a book, a table).

Objective

To collaboratively write several proverbs, one word at a time.

Instructions

1. Once your group of four or five people is sitting in a circle, paper and pens at the ready, a group leader says "go."
2. On "go," everyone writes down the first word of an imaginary proverb, then passes the sheet of paper as quickly as possible to the person on their right (and receives a sheet from the person on their left).
3. Read the word you've received, quickly add a second that would follow it, then immediately pass the paper to the right.

   (For example, if you've received **Never**, you might add **trust** or **eat** or whatever—the key is not to think too hard and to keep the collaboration moving.)

4. Read the budding proverb you've received, then quickly add a new word and pass it on.

   (For example, if you've received **Snowflakes gather**, you might add **on** or **when**—small words are important, too!)

5. Continue writing and passing, one word at a time, until someone in your group considers a proverb "done." When that occurs, that person should read the sentence aloud.* The small group will then bask in its newly created wisdom.
6. Keep passing the other pages along until you've created several collaborative proverbs.
7. When all groups are finished, share one or two of your proverbs with the whole community. More basking in wisdom will follow.

   *Variation: When a proverb feels "done," some groups like to defer reading it aloud; instead, they just drop the finished proverb on the floor or table in front of them and continue receiving and passing the remaining sheets of paper. Only when all the proverbs are finished does the group read them aloud. This practice creates suspense, keeps the flow moving, and results in a high concentration of laughter/wisdom.

20

SOURCE STYLE SCRAMBLE

Brennan Thomas, Molly Fischer, and Jodi Kutzner

Objectives

- To familiarize tutors with the formatting guidelines of different citation styles used in various disciplines.
- To enhance tutors' knowledge and appreciation of the similarities and differences between citation styles.
- To increase tutors' ability to assist students with properly formatting source entries using their disciplines' preferred citation styles.

Number of players

Any number of tutors (peer or faculty) can play this game. Tutors can compete against each other individually or on teams, or a single player can compete against a timer.

Materials

- Per player (for player-against-player competitions) or per team (for team competitions): Ten handwritten or typed and printed source entries (formatted in APA, MLA, Chicago, or any other citation style), cut into individual components (e.g., author, source title, date of publication).
- Per player (for player-against-player competitions) or per team (for team competitions): Ten handwritten or typed and printed source entry labels indicating each entry's type and style (e.g., "Article from a scholarly journal—MLA").
- Scoring sheet.
- Writing utensil.
- Timer (optional).
- Three additional source entries and labels in case of a tie (optional).

Starting the Game/Setup

1. Source entries should be written or printed on paper and then cut into individual pieces (e.g., author's name, title, date of publication). Source entry labels listing type and citation style (e.g., "Print book with single author—APA") should be written or printed out as well. Each entry's

https://doi.org/10.7330/9781646421947.c020

pieces and label should be kept together (ideally, paper-clipped and stored in an envelope or folder) until starting the game.

2. Before starting the game, tutors should choose which game variation they would like to play: player against player, team against team, or single player against a timer. A non-playing tutor or faculty member should also serve as the game's referee to determine the winner of each round. (If a single player is competing against a timer, no referee is needed.)

Rules of Play

1. The game consists of ten rounds. To start the first round, each player or team is provided with the pieces (e.g., author, title, publication date) and source label (e.g., "Short work from a website—MLA") for one entry.
2. The player or team must then rearrange these pieces in the correct order according to the source label's type (e.g., "short work from a website") and style (e.g., MLA).
3. The first player or team to put all pieces in the correct order (as determined by the game's referee) earns one point. If two players or teams tie, both players or teams earn one point.
4. The referee records these points on the scoring sheet.
5. Each round proceeds in this fashion.
6. The game continues until all ten rounds have been played.

Winning/Completing the Game

The player or team with the most points at the end of the tenth round wins the game. If two or more players or teams have tied, these players or teams may continue with three additional rounds to determine the winner.

Playing Tips

- Prepare for the game by reviewing the formatting guidelines of designated citation styles (e.g., APA, MLA, Chicago).
- Encourage teams or individual players to consult style handbooks as needed during the game.
- Note any citation style guidelines that competitors seem unfamiliar with or have difficulty recalling.
- Review these guidelines with competitors after each game.
- For single players competing against timers rather than other tutors, record their times and award special prizes at the semester's end for top scores.

Rationale for the Game

In her groundbreaking work *Noise from the Writing Center*, Elizabeth H. Boquet (2002) contends that effective tutor training programs embolden tutors to take risks and (to quote Peter Elbow) "voyage out" of their comfort zones (79–80). Training workshops must do more than acquaint participants with the pedagogical theories underlying writing center work, therefore. They must enhance tutors' awareness of different disciplines' writing conventions and strengthen their abilities to assist writers with varying backgrounds, as noted by Sue Dinitz and Susanmarie Harrington (2014). Our writing center's training program achieves these objectives through fun, interactive games, from role-playing exercises and tutor simulations to team competitions and timed trials. Many of our games are designed to help tutors expand their knowledge of discipline-specific conventions and apply this knowledge to tutoring situations. Our *Source Style Scramble* is one such game.

Most writing center tutors are comfortable with the citation styles they use for their major coursework. Tutors majoring in English or philosophy, for instance, generally use MLA, while psychology and business majors prefer APA. However, because tutors often assist students with source documentation in citation styles they themselves do not use, they must become more familiar with these different styles' guidelines. This exercise reinforces these guidelines and enables tutors to recognize similarities and differences between their disciplines' styles and others. For instance, English majors who play this game will learn that for APA-formatted sources, authors' first names are abbreviated and ampersands (&) are used to separate listed authors' names instead of the word *and*. Likewise, psychology and education majors will see that for MLA-formatted source entries, titles rather than publication dates are listed after authors' names and all key words in titles are capitalized. This game offers a simple yet engaging way to familiarize tutors with these style variations and thus more ably assist students from all disciplines with source documentation and formatting.

REFERENCES

Boquet, Elizabeth H. 2002. *Noise from the Writing Center*. Logan: Utah State University Press.

Dinitz, Sue, and Susanmarie Harrington. 2014. "The Role of Disciplinary Expertise in Shaping Writing Tutorials." *Writing Center Journal* 33, no. 2: 73–98.

SUPPLEMENTAL MATERIALS

Below are several sample entries (first scrambled and then arranged in the correct order) in APA, MLA, and Chicago.

Journal article: APA

Pieces

Cathro, V., O'Kane, P., & Gilberston, D.
427–42.
(2017).
Assessing reflection.
doi:10.1108/ET-01-2017-0008
(4),
Education & Training,
59

In the correct order:

Cathro, V., O'Kane, P., & Gilberston, D. (2017). Assessing reflection. *Education & Training,* 59(4), 427–42. doi:10.1108/ET-01-2017-0008.

Book, single author: MLA

Pieces

Eliot, Marc.
2009.
Crown/Archetype,
American Rebel: The Life of Clint Eastwood.

In the correct order:

Eliot, Marc. *American Rebel: The Life of Clint Eastwood.* Crown/Archetype, 2009.

Book, single author: Chicago

Pieces

New York:
Parton, Nigel.
1985.
The Politics of Child Abuse.
St. Martin's Press,

In the correct order:

Parton, Nigel. *The Politics of Child Abuse.* New York: St. Martin's Press, 1985.

INDEX

ABOUT THE AUTHORS

Neil Baird is associate professor of English and director of the University Writing Program at Bowling Green State University. Before moving to BGSU, he was writing center director at Western Illinois University for ten years. He is currently analyzing data collected from a multi-year, interview-driven case study of writing transfer in the major. With Bradley Dilger, he has published in *College Composition and Communication*, *WPA: Writing Program Administration*, and *Across the Disciplines*. As a member of the Elon Research Seminar on Writing beyond the University, his next research project will focus on the writing transfer practices of alumni at work.

Brenta Blevins is assistant professor of writing studies and digital studies at the University of Mary Washington. She teaches and researches multimodal and digital literacy, rhetoric, and pedagogy. Her publications have addressed digital media and pedagogy, multimodality, and multiliteracy centers in *Computers and Composition*, *Peer Review*, and various collections. Her dissertation "From Corporeality to Virtual Reality: Theorizing Literacy, Bodies, and Technology in the Emerging Media of Virtual, Augmented, and Mixed Realities" received the Honorable Mention for the 2017 Computers and Composition Hugh Burns Best Dissertation Award.

Elizabeth Caravella is assistant professor of visual studies at York University. Her research focuses on gameful design and the role of procedural rhetoric in composition pedagogy, as well as how video games use multimodal spaces and visual cueing to craft embodied arguments that influence and alter player dispositions. In addition to this and other edited collections, her work has also appeared in *Computers and Composition* and *Technical Communication Quarterly*, and she is the member coordinator of the Council for Play and Game Studies.

Jessica Clements is associate professor of English and director of the Composition Commons at Whitworth University. She has served as style editor for *Present Tense* since 2012 and as managing editor since 2020. Her scholarship centers on ethos and the role of human and object-oriented actors in contemporary multimodal communication. She is collaborating on an interdisciplinary book evaluating the influence of social media networks in shaping binary-bound parenting decisions. She has published on the pedagogical performance of faith in the first-year writing classroom, as well as on writing center consultant continued education topics including, the role of new media expertise in shaping writing center consultations.

Jason Custer is assistant professor of English at Midway University Moorhead. His research focuses on digital multimodality, digital rhetoric, procedural rhetoric, writing centers, multimodal writing centers, games and learning, technical writing, and programmatic assessment. He tweets at @CusterTeaching.

Malcolm Evans holds bachelor's and master's degrees in communication studies from Kean University. His research interests include escape rooms and using play as pedagogy in higher education. During his time in Kean's writing center, he presented at both the

International Writing Centers Association and Mid-Atlantic Writing Centers Association Conferences. He is studying for his PhD in interpersonal communication at Louisiana State University.

Richonda Fegins is associate director of alumni and philanthropic communications at Seton Hall University. In her role, she focuses on writing and designing fundraising appeals and campaign solicitations to help build the university's donor base. Prior to this, Richonda worked at Kean's writing center for over four years, helping students become more confident about writing and guiding them to establish long-term proficiency. Richonda received both her bachelor's and master's degrees in English and writing studies from Kean University.

Molly Fischer is a 2019 graduate of Saint Francis University, earning BA degrees in both English and women's studies. She tutored in the university's writing center for four years, the entirety of her undergraduate career. Through the writing center she was able to present scholarship at the Mid-Atlantic Writing Centers Association Conferences in 2017 and 2019. She also completed Saint Francis University's honors program; for her senior honors thesis, she composed a full-length novel focusing on the personal struggles of a high school student approaching the end of her senior year. She is revising and editing this novel and hopes to publish it in the near future, as well as continue to write young adult fiction.

Elliott Freeman is a research and writing librarian at Louisiana State University Health Shreveport. His professional interests include writing center theory, game studies, pop culture, and poetry. His poetry has appeared in journals such as *Rust+Moth*, *Liminality*, and *Rogue Agent*; he also works as a freelance writer for tabletop role-playing games, including *Exalted* and *Scion*.

Veronica Garrison-Joyner is a PhD student at George Mason University focusing on multiliteracy pedagogy, digital rhetoric, technical communication, and social justice. Drawing from experiences with diverse student populations in various writing center contexts and current multiliteracy pedagogical theory, she looks at ways of shaping writing center interactions to support and enhance students' engagement with multiple modes and languages in academic, professional, and civic settings. Her present research in the rhetoric of health and medicine aims to apply a multiliteracy perspective to the historically grounded situation of racial health disparities.

Jamie Henthorn is assistant professor of English and writing center director at Catawba College. She teaches first-year writing, game design, digital writing, professional and technical writing, and visual rhetoric. She writes about and has published on games, fitness, and geek culture from a cultural rhetorical perspective. On Twitter, you can find her thoughts on the X-Men and pictures of her cats at @JamieHenthorn.

Stacey Hoffer found her passion for writing center scholarship as an undergraduate tutor and writing fellow for the University of Delaware English department, where she taught workshops and conducted research on how to support writing and presentation skills for both bilingual and international students. She has been the writing center coordinator at Delaware Tech since fall 2013.

Jodi Kutzner graduated from Saint Francis University in 2018 with a BS in healthcare studies, as well as minors in Spanish–health sciences and biology. She tutored in the university's writing center from 2015 to 2018, mentoring many of the center's new hires during that interim. In addition to presenting scholarship at the Mid-Atlantic Writing Centers

Association Conferences in 2016 and 2017, Kutzner also participated in a three-member panel at the 2018 Mid-Atlantic Conference on College Composition and Communication (CCCC) hosted by Virginia Commonwealth University. She completed post-graduate studies in public health at the University of Pittsburgh and health behavior at Indiana University Bloomington.

Miriam E. Laufer serves as academic support specialist in the Learning Assistance Center at Howard Community College in Columbia, Maryland. She formerly served as manager of the Writing, Reading, and Language Center at Montgomery College, Germantown campus, and as president of the Maryland College Learning Centers Association (MDCLCA). She holds a master's degree in humanities from the University of Chicago. Her research interests include gamification in the writing center, the history of citations, and how feedback contributes to student success.

Christopher LeCluyse is professor of English and writing center director at Westminster College in Salt Lake City, Utah. His research relates medieval literacy, ancient rhetoric, religion, and now game studies to writing center praxis. He served as president of the Rocky Mountain Writing Centers Association and co-chaired the 2015 National Conference on Peer Tutoring in Writing and the 2017 International Writing Centers Association Summer Institute. His work has appeared in *Praxis*, *WLN: A Journal of Writing Center Scholarship*, and the edited collections *(E)Merging Identities: Graduate Students in the Writing Center* and *Writing Program Architecture: Thirty Cases for Reference and Research*.

Katie Levin (she/her/hers) is at the University of Minnesota–Twin Cities as co-director of the Center for Writing and affiliate graduate faculty member in literacy and rhetorical studies. She began her work in writing centers at Skidmore College and continued at Indiana University Bloomington, where she received her MA and PhD in English and wrote a qualitative dissertation study of a large WAC writing center. At Minnesota, in addition to consulting in the writing center, she co-designs and co-leads staff development activities; she also leads a research grants program and an annual Dissertation Writing Retreat. Her favorite writing is collaborative: she has coauthored pieces published in the *Writing Center Journal* and *Praxis: A Writing Center Journal*, as well as in a book from WAC Clearinghouse. Katie is particularly interested in collective work for racial and social justice in and through writing centers.

Christina Mastroeni completed her MA in writing studies at Kean University in Union, New Jersey. She also earned a BA focusing in communication and a minor in marketing from Holy Family University in Philadelphia, Pennsylvania. She teaches at Kean full-time as a lecturer of professional writing. Prior to this position in the field of education, she developed a marketing background from previous roles as a marketing recruiter and director of marketing.

Elysse T. Meredith is manager of the Writing, Reading, and Language Center at Montgomery College, Germantown campus. She holds a PhD in medieval studies (English literature, French literature, and art history) from the University of Edinburgh. As a professional writing tutor, her research interests encompass linguistic diversity, emotional labor, and multilingual composition.

Christopher L. Morrow is professor and department head of English and Languages at Western Illinois University Tarleton State University. Before Tarleton, he served as director of the University Writing Center. A Shakespearean and book historian by training, his research interests have recently focused around game studies and new media literature and adaptation. He is particularly interested in how games, digital forms, and concepts from game studies can offer different perspectives into and engagements with previously familiar texts, such as Shakespeare's plays, or forms, such as the writing center consultation. His essays have appeared in *Papers of the Bibliographical Society of America, Studies in English Literature: 1500–1900,* and *South Central Review.*

Mitchell Mulroy is a senior at Aquinas College, double majoring in history and English with a writing emphasis. He is a member of the on-campus writing center, works in the archives of the school library, and is president of the Aquinas College History Club. He also writes about video games at RPGFan.com as a news editor.

Alyssa Noch is a graduate of Aquinas College. She has a BA in English with a writing emphasis and a BA in history with a minor in women's studies. Before she graduated, she was a writing center consultant and president of the Writer's Guild on campus. In addition, she has presented her work at various events, most notably at the Michigan Writing Center Association Conference at the University of Michigan Flint, in 2018 where she presented a version of her chapter in this collection. While also at Aquinas, she was an intern with the Greater Grand Rapids Women's History Council. She is pursuing a master's degree in public history and an Archival Administration Certificate at Wayne State University.

Kevin J. Rutherford is a lecturer in the Writing Program at the University of California Santa Barbara. He has served as assistant director of the Howe Writing Initiative at Miami University, where he received his PhD, and as associate director of Campus Writing Programs at SUNY Cortland. His research interests include game studies, object-oriented rhetoric, and digital rhetoric. His work has appeared in the collection *Rhetoric, Through Everyday Things* and *Indie Games in the Digital Age.*

Holly Ryan is associate professor of English, composition coordinator, and writing center coordinator at Pennsylvania State University Berks. She teaches courses in the writing and digital media major, including a course in writing center theory and practice, while also directing the writing center. She is past president of the Mid-Atlantic Writing Centers Association and is a member of the International Writing Centers Association Executive Board. Her research explores power dynamics in tutor interactions, community-building practices, tutor education, and supporting the administrative work of writing center professionals. Her work has appeared in a range of journals including *Writing Center Journal, WLN: A Journal of Writing Center Scholarship,* and *Praxis: A Writing Center Journal.* She is managing editor for *Prompt: A Journal of Academic Writing Assignments.*

Lindsay A. Sabatino is associate professor of English and director of the writing center at Wagner College. Her research and pedagogical interests focus on writing center theory and practice, multimodal composition, digital rhetoric, writing tutor education, and faculty development. More specifically, her recent work explores composing in digital environments, the relocation of writing centers into libraries, online tutoring, gaming studies, and writing studio spaces. Along with Dr. Brian Fallon, she is co-editor of *Multimodal Composing: Strategies for Twenty-First-Century Writing Consultations.* She has also published in *Computers and Composition, Writing Lab Newsletter, Peer Review,* and *Praxis.*

Elizabeth Saur is lecturer in the Writing Program at the University of California Santa Barbara. Previously, she was writing center director at SUNY Cortland, assistant director of composition at Miami University, and writing center supervisor and director at California State University Fullerton. Her research interests center on composition pedagogy, writing program administration, and affect theory with a focus on teacher development practices. Her work on negotiating the emotional experiences of teaching composition can be found in *Writing Program Administration* and the collection *Standing at the Threshold: Working through Liminality in the Composition and Rhetoric TAship.*

Heather Shay is a visiting assistant professor of sociology at the University of South Carolina Beaufort. Previously, she was assistant professor of sociology at Lake Superior State University. Her specializations include social psychology and social inequality. Her dissertation research examined identity work among tabletop role-playing gamers and has been published in the *Journal of Contemporary Ethnography.* Her teaching interests include social psychology, social inequality, theory, culture, and qualitative methods.

Thomas "Buddy" Shay is writing coordinator for the Academic Enhancement Center at Georgia Gwinnett College. He has fifteen years' experience working in writing centers at the community college and university levels and has been a coordinator for most of those years. His creative work can be found in *Everyday Fiction, Wisteria Review,* and *Poetry Forum.*

Nathalie Singh-Corcoran is a service professor of English at West Virginia University where she directs the Eberly Writing Studio and teaches both graduate and undergraduate courses in writing pedagogy and professional writing and editing. She is a past president of the International Writing Centers Association. Her research interests include writing tutor education, writing center assessment, and contingent faculty concerns. Her scholarship has appeared in several edited collections and the *ADE Bulletin, College English, Composition Forum, Kairos,* and *WLN: A Journal of Writing Center Scholarship.*

Brennan Thomas is associate professor of English at Saint Francis University in Loretto, Pennsylvania, where she has been teaching full-time since 2010. Thomas teaches courses in introductory, argumentative, and novel writing and writing pedagogy, as well as a first-year seminar on the animated films of Disney and Pixar Studios. She directs the university's writing center, for which she trains and supervises all writing tutors. Thomas also tutors in the center several hours each week during the spring and fall semesters and edits the center's website content and quarterly newsletter, the *Write Times.* Her research interests include writing center administration and popular media studies.

Stephanie Vie is associate dean of the Outreach College at the University of Hawai'i at Mānoa. She is a former writing center director at Fort Lewis College in Durango, Colorado. She researches social media as well as games, with an interest in the intersections between the two. In particular, she focuses on issues of data mining and surveillance in online and app-based games. Her work has appeared in journals such as *First Monday, Computers and Composition, Technical Communication Quarterly,* and *Computers and Composition Online,* and she is the editor or co-editor of the books *The* Pokémon Go *Phenomenon, Social Writing/ Social Media,* and *e-Dentity.*

Rachael Zeleny is assistant professor of English and integrated arts at the University of Baltimore. She was formerly writing program director at Alvernia University where she led the development of the Writing Fellows Program. Her research interests lie in pedagogy, visual rhetoric, feminist rhetoric, and Victorian literature. She has published in journals such as *Rhetoric Review, Peitho, Journal of Interdisciplinary Studies,* and *Rhetorica.*